Ben Stacy Jerrik (Ed.)

Nau (Clothing Retailer)

Ben Stacy Jerrik (Ed.)

Nau (Clothing Retailer)

Portland, Oregon

Part Press

Imprint

Publisher:
Part Press is a trademark of
International Book Market Service Ltd., 17 Rue Meldrum, Beau Bassin, 1713-01 Mauritius
Email: info@bookmarketservice.com
Website: www.bookmarketservice.com

Published in 2012

Printed in: U.S.A., U.K., Germany. This book was not produced in Mauritius.

ISBN: 978-613-8-64246-6

Contents

Articles

References

Nau (clothing retailer)

Nau is an urban & outdoor apparel clothing company, founded 2005 and based in Portland, Oregon, United States with strong social and environmental missions.

The company's "green" mission includes the use of textiles developed from high quality sustainable technical fibers and fabrics in line with the Global Organic Textile Standard (GOTS) [1]. Nau also makes use of ethical wool [2] within its products. Furthermore all employees of Nau are provided free public transport and carbon offsetting tokens purchased for corporate travel and product *to* customer deliveries.[3]

Nau's core "social" mission involves a donation of 2% from each sale, on occasion 10%[4] to selected community partners (chosen by the customer upon purchase) e.g. Mercy Corps,[5] Nau utilises these funds to work with partners toward a lasting change.

On May 2, 2008, Nau announced that it was ceasing operations, primarily due to an inability to raise further capital. It has now been purchased by Horny Toad, Inc. and reopened for trade in October 2008[6] both online and through a number of dealer locations around the U.S. and Western Europe[7] [8] [9] [10]

References

[1] http://www.global-standard.org/

[2] http://zque.co.nz/

[3] http://www.nau.com/about/our-design/business-revisited.html

[4] http://www.csrwire.com/press_releases/16059-Nau-Makes-Gifts-Of-Change

[5] http://www.mercycorps.org/partners/nau

[6] http://www.hornytoad.com/about/doing-good/nau.html

[7] Leap Of Faith | Fast Company (http://www.fastcompany.com/magazine/116/features-leap-of-faith.html)

[8] Outdoor Gear Goes Green : All-Natural Technical Materials Hit the Mainstream (By Roddy Scheer) (http://www.emagazine.com/view/?3954)

[9] the power of 5% (https://www.nau.com/partnersforchange/the_power_of_5_.jsp)

[10] Sustainable clothier Nau pulls the plug (http://blog.oregonlive.com/breakingnews/2008/05/sustainable_clothier_nau_pulls.html)

External links

- "And Nau for Something Completely Different (http://outside.away.com/outside/gear/200702/nau-1.html)," Andrew Tilin, Outside Magazine, Feb. 2007
- Nau Official Web Site (http://www.nau.com/)

Portland, Oregon

Portland

— City —

City of Portland

Portland's skyline from the west, with Mount Hood on the left

Flag

Seal

Nickname(s): "Rose City", "Stumptown", "PDX"[1] See Nicknames of Portland, Oregon for a complete listing.

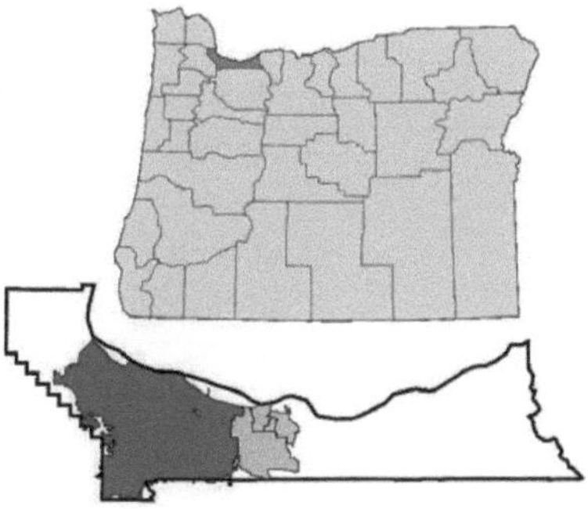

Location of Portland in Multnomah County and the state of Oregon

Portland

Location in the United States

Coordinates: 45°31′12″N 122°40′55″W

Country	USA
State	Oregon
Counties	Multnomah, Washington, Clackamas
Founded	1845
Incorporated	February 8, 1851
Government	
• Type	Commission
• Mayor	Sam Adams[2]
• Commissioners	Randy Leonard Dan Saltzman Nick Fish Amanda Fritz
• Auditor	LaVonne Griffin-Valade
Area	
• City	145.4 sq mi (376.5 km^2)
• Land	134.3 sq mi (347.9 km^2)
• Water	11.1 sq mi (28.6 km^2)
Elevation	50 ft (15.2 m)
Population (2010)	
• City	583,776
• Density	4288.38/sq mi (1655.31/km^2)
• Metro	2,260,000
• Demonym	Portlander
Time zone	PST (UTC-8)
• Summer (DST)	PDT (UTC-7)
ZIP codes	97086-97299
Area code(s)	503 and 971
FIPS code	41-59000[3]
GNIS feature ID	1136645[4]
Website	www.portlandonline.com [5]

Portland is a city located in the Pacific Northwest, near the confluence of the Willamette and Columbia rivers in the U.S. state of Oregon. As of the 2010 Census, it had a population of 583,776,[6] making it the 29th most populous city in the United States. Portland is Oregon's most populous city, and the third most populous city in the Pacific Northwest, after Seattle, Washington and Vancouver, British Columbia, Canada. Approximately 2,260,000 people live in the Portland metropolitan area (MSA),[7] the 23rd most populous in the United States.[7]

Portland was incorporated in 1851 and is the county seat of Multnomah County.[8] The city extends west into the Cedar Mill neighborhood in Washington County and south towards Lake Oswego in Clackamas County. With a commission-based government headed by a mayor and four other commissioners, the city and region are noted for strong land-use planning[9] and investment in light rail, supported by Metro, a distinctive regional government.

Because of its public transportation networks and efficient land use planning, Portland has been referred to as one of the most environmentally friendly, or "green", cities in the world.[10]

Located in the Marine west coast climate region, Portland has a climate marked by warm, dry summers and wet but mild winters. This climate is ideal for growing roses, and for more than a century, Portland has been known as "The City of Roses"[11] [12] with many rose gardens—most prominently the International Rose Test Garden. The city is also known for its large number of microbreweries and microdistilleries, as well as its coffee enthusiasm. It is also the home of the Timbers MLS team and the Trail Blazers NBA team.

History

Portland in 1890

The land today occupied by Multnomah County was inhabited for centuries by two bands of Upper Chinook Indians. The Multnomah people settled on and around Sauvie Island and the Cascades Indians settled along the Columbia Gorge. These groups fished and traded along the river and gathered berries, wapato and other root vegetables. The nearby Tualatin Plains provided prime hunting grounds.[13] The later settlement of Portland started as a spot known as "the clearing,"[14] which was on the banks of the Willamette about halfway between Oregon City and Fort Vancouver. In 1843, William Overton saw great commercial potential for this land but lacked the funds required to file a land claim. He struck a bargain with his partner, Asa Lovejoy of Boston, Massachusetts: for 25¢, Overton would share his claim to the 640 acres (2.6 km^2) site. Overton later sold his half of the claim to Francis W. Pettygrove of Portland, Maine. Pettygrove and Lovejoy each wished to name the new city after his respective home town. In 1845, this controversy was settled with a coin toss, which Pettygrove won in a series of two out of three tosses.[15] The coin used for this decision, now known as the Portland Penny, is on display in the headquarters of the Oregon Historical Society.

At the time of its incorporation on February 8, 1851 Portland had over 800 inhabitants,[16] a steam sawmill, a log cabin hotel, and a newspaper, the *Weekly Oregonian*. By 1879, the population had grown to 17,500.[17] The city merged with Albina and East Portland in 1891, and annexed the cities of Linnton and St. Johns in 1915.

Portland's location, with access both to the Pacific Ocean via the Willamette and the Columbia rivers and to the agricultural Tualatin Valley via the "Great Plank Road" through a canyon in the West Hills (the route of current-day U.S. Route 26), gave it an advantage over nearby ports, and it grew very quickly.[18] It remained the major port in the Pacific Northwest for much of the 19th century, until the 1890s, when Seattle's deepwater harbor was connected to the rest of the mainland by rail, affording an inland route without the treacherous navigation of the Columbia River.

Nicknames

The most common nickname for Portland is *The City of Roses,*[19] and this became the city's official nickname in 2003.[20] Other nicknames include the City of Bridges, *Stumptown,*[21] *Bridgetown,*[22] *Rip City,*[23] *Little Beirut,*[1] *Beervana*[24] [25] or *Beertown,*[26] *P-Town,*[20] [27] *Soccer City USA,*[28] [29] [30] [31] *Portlandia*, and the metonymous *PDX.*

Geography

The Willamette River runs through the center of the city, while Mount Tabor (center) rises on the city's east side. Mount Saint Helens (left) and Mount Hood (right center) are visible from many places in the city.

Topography

Portland lies at the northern end of Oregon's most populated region, the Willamette Valley. However, as the metropolitan area is culturally and politically distinct from the rest of the valley, local usage often excludes Portland from the valley proper. Although almost all of Portland lies within Multnomah County, small portions of the city lie within Clackamas and Washington counties with mid-2005 populations estimated at 785 and 1,455, respectively. The Willamette River runs north through the city center, separating the east and west sections of the city before veering northwest to join with the Columbia River (which separates the state of Washington from the state of Oregon) a short distance north of the city.

According to the United States Census Bureau, the city has a total area of 145.4 square miles (377 km^2). 134.3 square miles (348 km^2) of it is land and 11.1 square miles (29 km^2), or 7.6%, is water.[32]

Portland lies on top of an extinct Plio-Pleistocene volcanic field known as the Boring Lava Field.[33] The Boring Lava Field includes at least 32 cinder cones such as Mount Tabor,[34] and its center lies in Southeast Portland. The dormant but potentially active volcano Mount Hood to the east of Portland is easily visible from much of the city during clear weather. The active volcano Mount Saint Helens to the north in Washington is visible in the distance from high-elevation locations in the city and is close enough to have dusted the city with volcanic ash after an eruption on May 18, 1980.[35] Mount Adams, another prominent volcano in Washington state to the northeast of Portland, is also visible from parts of the city.

Climate

Portland experiences a temperate climate that is usually described as oceanic with mild, damp winters and relatively dry, warm summers. Like much of the Pacific Northwest, according to the Köppen climate classification it falls within the cool, dry-summer subtropical zone (*Csb*), also referred to as cool-summer Mediterranean, because of its relatively dry summers.[36] Other climate classification systems, such as Trewartha, place it firmly in the Oceanic zone (*Do*).[37]

Summers in Portland are warm, sunny and rather dry, with August, the warmest month, having an average daily high temperature of 79.7 °F (26.5 °C), and much larger day-night variation than in winter. Because of its inland location and when there is an absence of a sea breeze, heatwaves occur (in particular during the months of July and August) with air temperatures sometimes rising over 100 °F (38 °C), but 90 °F (32 °C) is more commonplace, occurring 13 days per annum.[38] Winters are normally mild, and very moist, with January averaging 39.9 °F (4.4 °C). Lows, though usually above freezing, can reach that mark or below 37 nights per year, however.[38] Cold snaps are short-lived, and snowfall occurs no more than a few times per year, although the city has been known to see major snow and ice storms because of the cold air outflow from the Columbia River Gorge. The city's winter snowfall totals have ranged from just a trace on many occasions, to 60.9 inches (154.7 cm) in 1892–93. Spring can bring rather unpredictable weather, resulting from warm spells, to thunderstorms rolling off the Cascade Range. The rainfall averages an equivalent 37.5 inches (950 mm) per year in downtown Portland, spread over 155 days a year. The lowest temperature ever recorded in Portland was , set on February 2, 1950.[39] The highest temperature ever recorded was 107 °F (42 °C), set on July 30, 1965 as well as August 8 and 10, 1981.[39] Temperatures of 100 °F (38 °C) have been recorded in each of the months from May through September.[39]

Cityscape

Panorama of downtown Portland. Hawthorne Bridge viewed from a dock on the Willamette River near the Oregon Museum of Science and Industry.

Panorama of downtown Portland at night. Viewed from across the Willamette River in SE Portland.

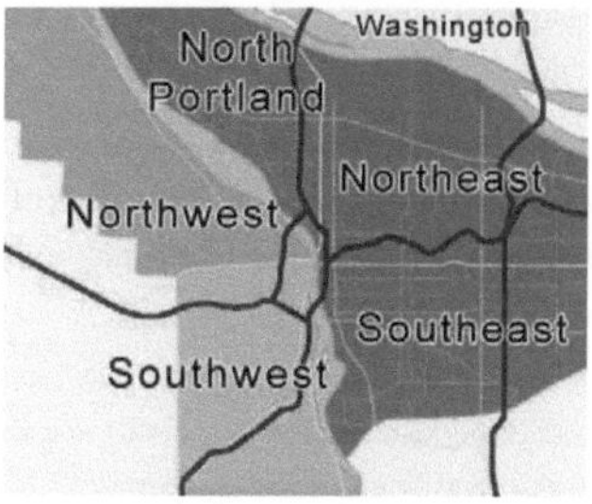

The "Five Quadrants" of Portland

Portland straddles the Willamette River near its confluence with the Columbia River. The denser and earlier-developed west side is mostly hemmed in by the nearby West Hills (Tualatin Mountains), though it extends over them to the border with Washington County. The flatter east side fans out for about 180 blocks, until it meets the suburb of Gresham. Rural Multnomah County lies farther east. In 1891 the cities of Portland, Albina, and East Portland were consolidated, and duplicate street names were given new names. The "great renumbering" on September 2, 1931 standardized street naming patterns, and changed house numbers from 20 per block to 100 per block. It divided Portland into five sections: Southwest, Southeast, Northwest, North, and Northeast. Burnside St. divides north and south, and the Willamette River divides east and west. The river curves west five blocks north of Burnside and in place of it, Williams Ave. is used as a divider. The North section lies between Williams Ave. and the Willamette River to the west.

On the west side, the RiverPlace, John's Landing and South Waterfront Districts lie in a "sixth quadrant" where addresses go higher from west to east toward the river. This "sixth quadrant" is roughly bounded by Naito Parkway and Barbur Boulevard to the west, Montgomery Street to the north and Nevada Street to the south. East-West addresses in this area are denoted with a leading zero (instead of a minus sign). This means 0246 SW California St. is not the same as 246 SW California St. Many mapping programs are unable to distinguish between these two different addresses.

Parks and gardens

A panoramic view of the International Rose Test Garden

Tom McCall Waterfront Park seen from the north

Portlanders are proud of Portland's parks and its legacy of preserving open spaces. Parks and greenspace planning date back to John Charles Olmsted's 1903 *Report to the Portland Park Board.* In 1995, voters in the Portland metropolitan region passed a regional bond measure to acquire valuable natural areas for fish, wildlife, and people. Ten years later, more than 8100 acres (33 km^2) of ecologically valuable natural areas had been purchased and permanently protected from development.[42]

Portland is one of only three cities in the contiguous U.S. with extinct volcanoes within its boundaries (besides Jackson, Mississippi and Bend, Oregon). Mount Tabor Park is known for its scenic views and historic reservoirs.[43]

Forest Park is the largest wilderness park within city limits in the United States, covering more than 5000 acres (2023 ha). Portland is also home to Mill Ends Park, the world's smallest park (a two-foot-diameter circle, the park's area is only about 0.3 square m). Washington Park is just west of downtown, and is home to the Oregon Zoo, the Portland Japanese Garden, and the International Rose Test Garden.

Tom McCall Waterfront Park runs along the west bank of the Willamette for the length of downtown. The 37-acre (15 ha) park was built in 1974 after Harbor Drive was removed and now hosts large events throughout the year. Portland's downtown features two groups of contiguous city blocks dedicated for park space: the North and South Park Blocks.

Tryon Creek State Natural Area is one of three Oregon State Parks in Portland and the most popular; its creek has a run of steelhead. The other two State Parks are Willamette Stone State Heritage Site located in the West Hills and the Government Island State Recreation Area located in the Columbia River near Portland International Airport.

Culture and contemporary life

Portland is often awarded the "Greenest City in America", and ranks among the world's top 10 greenest cities. *Popular Science* has continued to award Portland the title of the Greenest City in America and *Grist* magazine lists it as the second greenest city in the world.[44] [45] Portland is the home city of "the World's Oldest Teenage Drag Queen Pageant", the Rose Bud and Thorn Pageant, started in 1975 and modeled after the Imperial Sovereign Rose Court of Oregon.[46]

Entertainment and performing arts

Like most large cities, Portland has a range of classical performing arts institutions which include the Oregon Ballet Theatre, Oregon Symphony, Portland Opera and the Portland Youth Philharmonic. It also has quite a few stages similar to New York's Off Broadway or Off-Off-Broadway such as Portland Center Stage, Artists Repertory Theatre, Miracle Theatre, Stark Raving Theatre, and Tears of Joy Theatre. Portland hosts the world's only HP Lovecraft Film Festival[47] at the Hollywood Theatre.

The Arlene Schnitzer Concert Hall, home of the Oregon Symphony, among others

Portland is home to famous bands such as The Kingsmen and Paul Revere & the Raiders, both famous for their associations with the song "Louie Louie". Other widely known musical groups include The Dandy Warhols, Everclear, Modest Mouse, Pink Martini, Sleater-Kinney, The Shins, Blitzen Trapper, The Decemberists, and the late Elliott Smith. The city's now-demolished Satyricon nightclub is well known for being the place where the late Nirvana frontman Kurt Cobain and Hole frontwoman Courtney Love met each other; Love had grown up in Portland for most of her life.[48] In recent years, a number of indie music bands from Portland have been touring nationally.

Widely recognized animators who hail from Portland include Matt Groening (*The Simpsons*) , Will Vinton (Will Vinton's *A Claymation Christmas Celebration*) and Travis Knight (*Coraline*). Dan Steffan, cartoonist-illustrator for *Heavy Metal* and other magazines, lives in Portland.

Filmmaker Gus Van Sant (*Good Will Hunting* (1997), *Milk* (2008)) is also a Portland native. Actors from Portland include Sam Elliott and Sally Struthers.

Recent films set and shot in Portland include *Extraordinary Measures*, *Body of Evidence*, *What the Bleep Do We Know!?*,*The Hunted*, *Twilight*, *Paranoid Park*, *Wendy and Lucy*, *Feast of Love*, *Untraceable*, and *Coraline* . An unusual feature of Portland entertainment is the large number of movie theaters serving beer, often with second-run or revival films. A notable example of these "brew and view" theaters is The Bagdad Theater and Pub.

TV shows including *Portlandia*, *Leverage*, *Under Suspicion*, *Grimm*, *Nowhere Man* and *Life Unexpected* have been filmed in Portland.

Authors

Authors from Portland include science fiction writer Ursula K. Le Guin, famous for her *Earthsea* novels, *Hainish Cycle* and *Orsinian Tales*; transgressional fiction novelist Chuck Palahniuk, best known for his award-winning novel *Fight Club*; best-selling Christian author Don Miller; and Beverly Cleary, author of the famous series of children's books featuring Henry Huggins, his dog Ribsy, Beatrice "Beezus" Quimby and Ramona Quimby. Klickitat Street, where Cleary's characters live, is an actual street in northeast Portland. Statues of the characters stand in nearby Grant Park.

Portland is home to a number of independent, small graphic novel publishers such as Dark Horse Comics and Oni Press,[49] as well as comic book artists and writers such as Brian Michael Bendis and Farel Dalrymple.

Tourism

The White Stag sign is a popular city landmark

Portland is home to a diverse array of artists and arts organizations, and was named in 2006 by *American Style* magazine as the tenth best Big City Arts Destination in the U.S.

The Portland Art Museum owns the city's largest art collection and presents a variety of touring exhibitions each year. With the recent addition of the Modern and Contemporary Art wing it became one of the United States' twenty-five largest museums. Art galleries abound downtown and in the Pearl District, as well as in the Alberta Arts District and other neighborhoods throughout the city.

The Oregon Museum of Science and Industry (OMSI) is located on the east bank of the Willamette River across from downtown Portland, and contains a variety of hands-on exhibits covering the physical sciences, life science, earth science, technology, astronomy, and early childhood education. OMSI also has an OMNIMAX Theater and is home to the USS *Blueback* submarine, used in the film *The Hunt for Red October*.

Portland is also home to Portland Classical Chinese Garden, an authentic representation of a Suzhou-style walled garden.

Portlandia, a statue on the west side of the Portland Building, is the second-largest hammered-copper statue in the U.S. (after the Statue of Liberty). Portland's public art is managed by the Regional Arts & Culture Council.

Powell's City of Books, whose downtown multistory building at the intersection of Burnside and 10th Street occupies an entire city block, claims to be the largest independent bookstore in the United States and the largest bookstore west of the Mississippi River. Their specialty computer and science bookstore recently moved into a new location across the street.

The Portland Rose Festival takes place annually in June and includes two parades, dragon boat races, carnival rides at Tom McCall Waterfront park, and dozens of other events.

Washington Park, in the West Hills, is home to some of Portland's most popular recreational sites, including the Oregon Zoo, the Portland Japanese Garden, the World Forestry Center, and the Hoyt Arboretum.

Portland hosts a number of festivals throughout the year in celebration of beer and brewing, including the Oregon Brewers Festival. Held each summer during the last full weekend of July, it is the largest outdoor craft beer festival in North America with over 70,000 attendees in 2008.[50] Other major beer festivals throughout the calendar year include the Spring Beer and Wine Festival in April, the North American Organic Brewers Festival in June, the Portland International Beerfest in July,[51] and the Holiday Ale Festival in December.

Shopping

Portland has many options for shopping. Some of the well known shopping areas are Downtown Portland, Nob Hill (NW 21st & 23rd Avenues), Pearl District, and the Lloyd District. Major department stores in downtown include Nordstrom, Macy's, and Mario's. The major malls in the metropolitan area are Bridgeport Village, Washington Square, Clackamas Town Center, Lloyd Center, Vancouver Mall, and Pioneer Place. Another destination is the Portland Saturday Market, a town bazaar-like environment where many kinds of goods are sold from Artisan Crafts to Tibetan Imports, reflecting the many cultures of Portland. The Saturday Market is open every weekend from March through Christmas.

Hawthorne District

Breweries

Portland is well known for its microbrewery beer.[52] Oregon Public Broadcasting has documented Portland's role in the microbrew revolution in the United States in a report called *Beervana*.[53] Some illustrate Portlanders' interest in the beverage by an offer made in 1888 when local brewer Henry Weinhard volunteered to pump beer from his brewery into the newly dedicated Skidmore Fountain. Portland's modern abundance of microbreweries dates to the 1980s when state law was changed to allow consumption of beer on brewery premises. Brewery innovation was supported by the abundance of local ingredients, including two-row barley, over a dozen varieties of hops, and pure water from the Bull Run Watershed.

Portland is home to more than 40 breweries—more breweries than any other city in the world[54] —which is partially responsible for CNBC naming Portland the best city for happy hour in the U.S.[55] The McMenamin brothers alone have over thirty brewpubs, distilleries, and wineries scattered throughout the metropolitan area, several in renovated cinemas and other historically significant buildings otherwise destined for demolition. Other notable Portland brewers include Widmer Brothers, BridgePort, and Hair of the Dog, as well as numerous smaller, quality brewers. In 1999, author Michael "Beerhunter" Jackson called Portland a candidate for the *beer capital of the world* because the city boasted more breweries than Cologne, Germany. The Portland Oregon Visitors Association promotes "Beervana" and "Brewtopia" as nicknames for the city.[56] In mid-January 2006, Mayor Tom Potter officially gave the city a new nickname: Beertown.[57]

Cuisine

Portland has a growing restaurant scene, and among three nominees, was recognized by the Food Network Awards as their "Delicious Destination of the Year: A rising city with a fast-growing food scene" for 2007.[58]

The original Stumptown Coffee location at 45th and Division.

The *New York Times* also spotlighted Portland for its burgeoning restaurant scene in the same year.[59] *Travel + Leisure* ranked Portland No. 9 among all national cities in 2007.[60] The city is also known for being the most vegetarian-friendly city in America.[61]

Portland has been named the best city in the world for street food by several publications, including U.S. News[62] and CNN.[63] Food cart pods, approaching 500 in 2010[64] , spread throughout the city have come to define the scene, appearing frequently on television programs & becoming popular destinations themselves.

In addition to beer, Portland has become known as a premier coffee destination in the Pacific Northwest. Yelp.com lists more than 20 coffee houses in Portland with 4.5-5 star ratings.[65] The city is home to the original Stumptown Coffee Roasters, well known by aficionados as one of the nation's highest quality direct-trade roasteries,[66] as well as dozens of other micro-roasteries and cafes.

Sports

Portland is home to two major league teams: the Portland Timbers of Major League Soccer and the Portland Trail Blazers of the National Basketball Association.[67] The city is also home to a number of minor league teams.

The Rose Garden, home of the Portland Trail Blazers.

Running is a popular sport in the metropolitan area, which hosts the Portland Marathon and much of the Hood to Coast Relay, the world's largest (by number of participants) long-distance relay race. The city is home to the Oregon Track Club, which includes American record holders Chris Solinsky, Alan Webb, and Shalane Flanagan. Skiing and snowboarding are also highly popular, with a number of nearby resorts on Mount Hood, including year-round Timberline Lodge.

Portland was formerly home to the Portland Rosebuds of the Pacific Coast Hockey Association, the first professional sports team in Oregon and the first professional hockey team in the United States. The Rosebuds played in the 1916 Stanley Cup Finals, the first American team to do so.

The city also has one of the most active bicycle racing scenes in the United States. The Oregon Bicycle Racing Association sanctions hundreds of bicycling events each year. Weekly events at Alpenrose Velodrome and Portland International Raceway allow for racing nearly every night of the week March through September, and cyclocross races September through December, such as the Cross Crusade, can have over 1,000 riders and boisterous spectators.

Additionally, the Portland metro area has its own cricket league, the Oregon Cricket League. The league hosts two formats.[68] [69]

Portland has two Division I college sports teams, the University of Portland Pilots and the Portland State Vikings. Both universities field teams in numerous sports, including soccer, baseball, basketball, and football. The University of Portland plays at Joe Etzel Field, the Clive Charles Soccer Complex, and the Chiles Center. Portland State University plays at the Stott Center and Jeld-Wen Field. In addition, Lewis & Clark College fields several sports teams that compete in NCAA Division III.

Club	Sport	League	League championships	Home venue	Founded
Portland Trail Blazers	Basketball	National Basketball Association	1 (1976-77)	Rose Garden	1970
Portland Timbers	Soccer	Major League Soccer	0	Jeld-Wen Field	1975
Portland Winterhawks	Ice hockey	Western Hockey League	2 (1982–83, 1997–98)	Rose Garden	1976

Media

The Oregonian is the only daily general-interest newspaper serving Portland. It also circulates throughout the state and in Clark County, Washington.

Smaller local newspapers, distributed free of charge in newspaper boxes and at venues around the city, include *Portland Tribune* (general-interest paper published on Thursdays), *Willamette Week* (general-interest alternative weekly published on Wednesdays), *The Portland Mercury* (another weekly, targeted at younger urban readers published on Thursdays), and *The Asian Reporter* (a weekly covering Asian news, both international and local).

Portland Indymedia is one of the oldest and largest Independent Media Centers. The *Portland Alliance*, a largely anti-authoritarian progressive monthly, is the largest radical print paper in the city. *Just Out*, published in Portland twice monthly until the end of 2011, was the region's foremost LGBT publication. A biweekly paper, *Street Roots*, is also sold within the city by members of the homeless community.

The Portland Business Journal, a weekly, covers business-related news, as does *The Daily Journal of Commerce*. *Portland Monthly* is a monthly news and culture magazine. *The Bee*, over 105 years old, is another neighborhood newspaper serving the inner southeast neighborhoods.

Portland is well served by television and radio.

Economy

The Portland House-Price Index has remained stronger than the national average.

Portland's location is beneficial for several industries. Relatively low energy cost, accessible resources, North-South and East-West Interstates, international air terminals, large marine shipping facilities, and both west coast intercontinental railroads are all economic advantages.[70] The US consulting firm Mercer, in a 2009 assessment "conducted to help governments and major companies place employees on international assignments", ranked Portland 42nd worldwide in quality of living; the survey factored in political stability, personal freedom, sanitation, crime, housing, the natural environment, recreation, banking facilities, availability of consumer goods, education, and public services including transportation.[71]

The city's history of attracting and retaining company headquarters is mixed. Major businesses such as Willamette Industries, Louisiana-Pacific, CH2M HILL, U.S. Bank, and Evraz North American (formerly known as Oregon Steel Mills), have moved headquarters out of the city, as have smaller companies such as Lucy Activewear and Northwest Pipe Company.[72] Examples of how the city has attracted a company's world, North American, or U.S. headquarters include Vestas Wind Systems, and sporting goods manufacturers Li-Ning Co., Hi-Tec Sports, KEEN, Inc. and Adidas.[72]

Other Portland based companies include advertising firm Wieden+Kennedy; financial services companies Umpqua Holdings Corporation and StanCorp Financial Group; data tracking firm Rentrak; utility providers PacifiCorp, NW Natural and Portland General Electric; communications provider Integra Telecom; restaurant chains McMenamins and McCormick & Schmick's; toolmaker Leatherman; and architectural firms ZGF Architects LLP and Boora Architects.

Real estate and construction

Oregon's 1973 "urban growth boundary" law limits the boundaries for large scale development in each metropolitan area in Oregon.[73] This limits access to utilities such as sewage, water and telecommunications, as well as coverage by fire, police and schools.[73] Originally this law mandated that the city must maintain enough land within the boundary to provide an estimated 20 years of growth; however, in 2007 the legislature altered the law to require the maintenance of an estimated 50 years of growth within the boundary, as well as the protection of accompanying farm and rural lands.[74]

The growth boundary, along with efforts of the PDC to create economic development zones, has led to the development of a large portion of downtown, a large number of mid- and high-rise developments, and an overall increase in housing and business density.[75] [76]

Manufacturing

Computer components manufacturer Intel is the Portland area's largest employer, providing jobs for more than 15,000 people, with several campuses to the west of central Portland in the city of Hillsboro.[70] The metro area is home to more than 1,200 technology companies.[70] This high density of technology companies has led to the nickname Silicon Forest being used to describe the Portland area, a reference to the abundance of trees in the region and to the Silicon Valley region in Northern California.

Portland is home to the North American headquarters for Adidas, while the metropolitan area serves as the headquarters for Nike, FLIR Systems, Columbia Sportswear, and TriQuint Semiconductor, among others. Nike and Portland based Precision Castparts are the only two Fortune 500 companies headquartered in Oregon. Other manufacturing companies based in Portland include Freightliner Trucks, Zidell Companies, The Collins Companies, while Western Star Trucks builds their trucks in the city.

The steel industry's history in Portland predates World War II. By the 1950s, the steel industry became the city's number one industry for employment.[77] The steel industry thrives in the region, with Schnitzer Steel Industries, a prominent steel company, shipping a record 1.15 billion tons of scrap metal to Asia during 2003.[77] Other heavy industry companies include ESCO Corporation and Oregon Steel Mills.

Logistics

Portland is the largest shipper of wheat in the United States,[78] [79] and is the second largest port for wheat in the world.[80] The marine terminals alone handle over 13 million tons of cargo per year, and is home to one of the largest commercial dry docks in the country.[81] [82] The Port of Portland is the third largest export tonnage port on the west coast of the U.S., though it is located about 80 miles (130 km) upriver.[70] [82]

Transportation

Union Station

MAX Light Rail is the centerpiece of the city's public transportation system

Portland Streetcar runs north-south through Downtown

Portland Aerial Tram connects the South Waterfront district with OHSU

The Portland metropolitan area has transportation services common to major U.S. cities, though Oregon's emphasis on proactive land-use planning and transit-oriented development within the urban growth boundary means that commuters have multiple well-developed options. A 2011 study by Walk Score ranked Portland 12th most walkable of fifty largest U.S. cities.[83] Some Portlanders use mass transit for their daily commute. In 2008, 12.6% of all commutes in Portland were on public transit.[84] TriMet operates most of the region's buses and the MAX (short for Metropolitan Area Express) light rail system, which connects the city and suburbs. Westside Express Service, or WES, opened in February 2009 as commuter rail for Portland's western suburbs, linking Beaverton and Wilsonville. The Portland Streetcar operates from the south waterfront, through Portland State University and north to nearby homes and shopping districts. The Streetcar line will add 3.3 miles of tracks on the east side of the Willamette River in the fall of 2012.[85] The line will complete a loop to the tracks on the west side of the river once the new Portland-Milwaukie Light Rail Bridge is completed in 2013.[86] Within the Free Rail Zone, a designated geographic area centered in downtown, rides on TriMet's MAX and streetcar systems are free. Fifth and Sixth avenues within downtown comprise the Portland Transit Mall, two streets devoted primarily to bus and light rail traffic with limited automobile access. Intense public transit development continues as two light rail lines are under construction, as well as a new downtown transit mall linking several transit options. TriMet also provides real-time tracking of buses and trains with its TransitTracker and even makes the data available to software developers so they can create customized tools of their own.[87]

I-5 connects Portland with the Willamette Valley, Southern Oregon, and California to the south and with Washington to the north. I-405 forms a loop with I-5 around the central downtown area of the city and I-205 is a loop freeway route on the east side which connects to the Portland International Airport. US 26 supports commuting within the metro area and continues to the Pacific Ocean westward and Mount Hood and Central Oregon eastward. US 30 has a main, bypass, and business route through the city extending to Astoria, Oregon to the west; through Gresham, Oregon, and the eastern exurbs, and connects to I-84, traveling towards Boise, Idaho.

Portland's main airport is Portland International Airport, located about 20 minutes by car (40 minutes by MAX) northeast of downtown. In addition Portland is home to Oregon's only public use heliport, the Portland Downtown Heliport. Amtrak, the national passenger rail system, provides service to Portland at Union Station on three routes. Long-haul train routes include the Coast Starlight (with service from Los Angeles to Seattle) and the Empire Builder (with service from Portland to Chicago.) The Amtrak Cascades commuter trains operate between Vancouver, British Columbia and Eugene, Oregon, and serve Portland several times daily.

Portland is the only city in the United States that owns operating mainline steam locomotives, donated to the city in 1958 by the railroads that ran them.[88] Spokane, Portland & Seattle 700 and the world-famous Southern Pacific 4449 can be seen several times a year pulling a special excursion train, either locally or on an extended trip. The "Holiday Express", pulled over the tracks of the Oregon Pacific Railroad on weekends in December, has become a Portland tradition over its seven years running.[89] These trains and others are operated by volunteers of the Oregon Rail Heritage Foundation, an amalgamation of rail preservation groups which is currently constructing a new engine house and rail heritage center adjacent to OMSI.[90]

The city is particularly supportive of urban bicycling and has been recognized by the League of American Bicyclists among others for its network of on street bicycling facilities and other bicycle-friendly services.[91] It ranks highly among the most bicycle friendly cities in the world.[92] The Bicycle Transportation Alliance sponsors an annual Bicycle Commute Challenge, in which thousands of commuters compete for prizes and recognition based on the length and frequency of their commutes.[93] Approximately 8% of commuters bike to work, the highest proportion of any major U.S. city and about 10 times the national average.[94]

Car sharing through Zipcar and U Car Share is available to residents of the city and some inner suburbs. Portland has a commuter aerial cableway, the Portland Aerial Tram, which connects the South Waterfront district on the Willamette River to the Oregon Health & Science University campus on Marquam Hill above.

Portland has five indoor skateparks and is home to historically significant Burnside Skatepark. Gabriel Skatepark is the most recent, which opened on July 12, 2008. Another fourteen are in the works.[95] The Wall Street Journal stated Portland "may be the most skateboard-friendly town in America."[96]

Law and government

Portland City Hall

The city of Portland is governed by the Portland City Council, which includes the Mayor, four Commissioners, and an auditor. Each is elected citywide to serve a four year term. The auditor provides checks and balances in the commission form of government and accountability for the use of public resources. In addition, the auditor provides access to information and reports on various matters of city government.

The city's Office of Neighborhood Involvement serves as a conduit between city government and Portland's 95 officially recognized neighborhoods. Each neighborhood is represented by a volunteer-based neighborhood association which serves as a liaison between residents of the neighborhood and the city government. The city provides funding to neighborhood associations through seven district coalitions, each of which is a geographical groupings of several neighborhood associations. Most (but not all) neighborhood associations belong to one of these district coalitions.

Portland and its surrounding metropolitan area are served by Metro, the United States' only directly elected metropolitan planning organization. Metro's charter gives it responsibility for land use and transportation planning, solid waste management, and map development. Metro also owns and operates the Oregon Convention Center, Oregon Zoo, Portland Center for the Performing Arts, and Portland Metropolitan Exposition Center.

The Multnomah County government provides many services to the Portland area, as do Washington and Clackamas counties to the west and south.

Portland strongly favors the Democratic Party. Although local elections are nonpartisan, most of the city's elected officials are Democrats.

Portland's delegation to the Oregon Legislative Assembly is entirely Democratic. In the current 76th Oregon Legislative Assembly, which first convened in 2011, four state Senators represent Portland in the state Senate: Diane Rosenbaum (District 21), Chip Shields (District 22), Jackie Dingfelder (District 23), and Rod Monroe (District 24), all Democrats. Portland sends six Representatives to the state House of Representatives: Jules Bailey (District 42), Lew Frederick (District 43), Tina Kotek (District 44), Michael Dembrow (District 45), Ben Cannon (District 46), and Jefferson Smith (District 47), also all Democrats.

Portland's federal representation in the United States House of Representatives is split between three congressional districts. Most of the city is in the 3rd District, represented by Earl Blumenauer, who served on the city council from 1986 until his election to Congress in 1996. Most of the city west of the Willamette River is part of the 1st District, the seat for which has been vacant since the resignation of David Wu in mid-2011. A small portion of the city is in the 5th District, represented by Kurt Schrader. All three are Democrats. A Republican has not represented a significant portion of Portland in the U.S. House of Representatives since 1975. Both of Oregon's senators, Ron Wyden and Jeff Merkley, are from Portland and are also both Democrats.

Sam Adams, the current mayor of Portland, became the city's first openly gay mayor in 2009.[97] In 2004, 59.7 percent of Multnomah County voters cast ballots against Oregon Ballot Measure 36, which amended the Oregon Constitution to prohibit recognition of same-sex marriages. The measure passed with 56.6% of the statewide vote.

Multnomah County is one of two counties where a majority voted against the initiative; the other is Benton County, which includes Corvallis, home of Oregon State University.[98]

On April 28, 2005, Portland became the only city in the nation to withdraw from a Joint Terrorism Task Force.[99] [100]

Planning and development

Aerial view of central Portland

The city consulted with urban planners as far back as 1903. Development of Washington Park and one of the country's finest greenways, the 40 Mile Loop, which interconnects many of the city's parks, began.

Portland is often cited as an example of a city with strong land use planning controls;[9] This is largely the result of statewide land conservation policies adopted in 1973 under Governor Tom McCall, in particular the requirement for an urban growth boundary (UGB) for every city and metropolitan area. The opposite extreme, a city with few or no controls, is typically illustrated by Houston, Texas.[101] [102] [103] [104] [105]

1966 photo shows sawdust-fired power plant on the edge of Downtown that was removed to make way for dense residential development. High rises to left in background were early projects of the Portland Development Commission.

Portland's urban growth boundary, adopted in 1979, separates urban areas (where high-density development is encouraged and focused) from traditional farm land (where restrictions on non-agricultural development are very strict).[106] This was atypical in an era when automobile use led many areas to neglect their core cities in favor of development along interstate highways, in suburbs, and satellite cities.

The original state rules included a provision for expanding urban growth boundaries, but critics felt this wasn't being accomplished. In 1995, the State passed a law requiring cities to expand UGBs to provide enough undeveloped land for a 20 year supply of future housing at projected growth levels.[107]

The Portland Development Commission is a semi-public agency that plays a major role in downtown development; it was created by city voters in 1958 to serve as the city's urban renewal agency. It provides housing and economic development programs within the city, and works behind the scenes with major local developers to create large projects.

In the early 1960s, the PDC led the razing of a large Italian-Jewish neighborhood downtown, bounded roughly by the I-405 freeway, the Willamette River, 4th Avenue and Market street.

Mayor Neil Goldschmidt took office in 1972 as a proponent of bringing housing and the associated vitality back to the downtown area, which was seen as emptying out after 5 pm. The effort has had dramatic effects in the 30 years since, with many thousands of new housing units clustered in three areas: north of Portland State University (between the I-405 freeway, SW Broadway, and SW Taylor St.); the RiverPlace development along the waterfront under the Marquam (I-5) bridge; and most notably in the Pearl District (between I-405, Burnside St., NW Northrup St., and NW 9th Ave.).

The Urban Greenspaces Institute, housed in Portland State University Geography Department's Center for Mapping Research, promotes better integration of the built and natural environments. The institute works on urban park, trail, and natural areas planning issues, both at the local and regional levels.

In October 2009, the Portland City Council unanimously adopted a climate action plan that will cut the city's greenhouse gas emissions to 80% below 1990 levels by 2050.[108]

According to *Grist* magazine, Portland is the second most eco-friendly or "green" city in the world trailing only Reykjavík, Iceland.[109] In 2010, Move, Inc. placed Portland in its "top 10 greenest cities" list.[110] [111]

Free speech

Because of strong free speech protections of the Oregon Constitution upheld by the Oregon Supreme Court *Henry v. Oregon Constitution 1987* which specifically found that full nudity and lap dances in strip clubs are protected speech,[112] Portland is widely considered to have more strip clubs per capita than Las Vegas or San Francisco.[113] [114] [115] Portland has been titled as "Pornland" for its strip clubs, erotic massage parlors, and high rate of child sex trafficking.[116] [117] The term was heavily used in 2010, but the term was referenced by Chuck Palahniuk in 2003.[118]

A judge dismissed charges against a nude bicyclist in November 2008 on the grounds that the city's annual World Naked Bike Ride "was a well-established tradition in Portland. The first instance occurred sometime around 1999 and had less than 7 participants; at the time it was jokingly referred to as "critical ass" (a play on Critical Mass bike rides). Participants would 'purchase' a bike from a local chain department store and then return it the next morning. It used to take place at midnight and lasted until the participants were stopped/arrested. The prankster aspect of it came in when the arresting officers didn't want to touch the naked cyclists in order to arrest them.[119] The 2009 Naked Bike Ride occurred without significant incident.[120] City police managed traffic intersections.[121] There were an estimated 3,000 to 5,000 participants. In June 2010 Portland's World Naked Bike Ride had an estimated 13,000 people.[122] [123] [124]

A state law prohibiting publicly insulting a person in a way likely to provoke a violent response was tested in Portland and struck down unanimously by the State Supreme Court as violating protected free speech and being overly broad.[125]

Demographics

As of 2010, there are an estimated 583,776 people residing in the city, organized into 235,508 households. The population density is 4,375.2 people per square mile. There are 265,439 housing units at an average density of 1989.4 per square mile (1,236.3/km²). Population growth in Portland increased 10.3% between 2000 and 2010.[126] Population growth in the Portland metropolitan area has outpaced the national average during the last decade, with 2008 estimates showing an 80% chance of population growth in excess of 60% over the next 50 years.[74]

2010 Census data

The racial makeup of the city was 73.9% White (405,938), 8.8% Hispanic or Latino (of any race) (48,285), 7.8% Asian (42,785), 7.8% Black or African American (42,711), 2.8% Native American (15,523), 0.6% Pacific Islander (3,564), and 3.0% from other races (16,347).

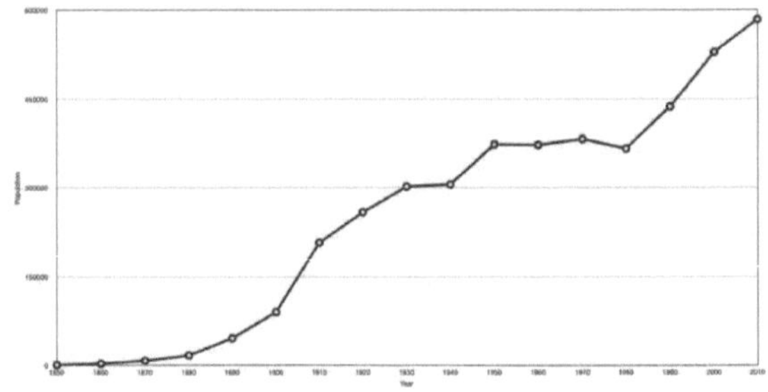

Graph showing the city's population growth from 1850 to 2010[6] [127] [128]

Out of 223,737 households, 24.5% have children under the age of 18 living with them, 38.1% are married couples living together, 10.8% have a female householder with no husband present, and 47.1% are non-families. 34.6% of all households are made up of individuals and 9% have someone living alone who is 65 years of age or older. The average household size is 2.3 and the average family size is 3.

The Marquam Bridge over the Willamette River, viewed from the southwest, atop Marquam Hill

The age distribution was 21.1% under the age of 18, 10.3% from 18 to 24, 34.7% from 25 to 44, 22.4% from 45 to 64, and 11.6% who are 65 years of age or older. The median age is 35 years. For every 100 females there are 97.8 males. For every 100 females age 18 and over, there are 95.9 males.

The median income for a household in the city is $40,146, and the median income for a family is $50,271. Males have a reported median income of $35,279 versus $29,344 reported for females. The per capita income for the city is $22,643. 13.1% of the population and 8.5% of families are below the poverty line. Out of the total population, 15.7% of those under the age of 18 and 10.4% of those 65 and older are living below the poverty line. Figures delineating the income levels based on race are not available at this time.

However, though the population of the city is increasing, the total population of children is diminishing, which has put pressure on the public school system to close schools. A 2005 study found that Portland is now educating fewer children than it did in 1925, despite the city's population having almost doubled since then, and the city will have to close the equivalent of three to four elementary schools each year for the next decade.[129]

In 1940, Portland's African-American population was approximately 2,000 and largely consisted of railroad employees and their families.[130] During the war-time liberty ship construction boom, the need for workers drew many blacks to the city. The new influx of blacks settled in specific neighborhoods, such as the Albina district and Vanport. The May 1948 flood which destroyed Vanport eliminated the only integrated neighborhood, and an influx of blacks into the NE quadrant of the city continued.[130] At 7.90%, Portland's African American population is nearly four times the state average. Over two thirds of Oregon's African-American residents live in Portland.[130] As of the 2000 census, three of its high schools (Cleveland, Lincoln and Wilson) were over 70% white, reflecting the overall population, while Jefferson High School was 87% non-white. The remaining six schools have a higher number of non-whites, including blacks and Asians. Hispanic students average from 3.3% at Wilson to 31% at Roosevelt.[131]

With about 12,000 Vietnamese residing in the city proper, Portland has one of the largest Vietnamese populations in America per capita.[132] According to statistics there are 21,000 Pacific Islanders in Portland, making up 4% of the population.[133]

The city of Portland has the 7th highest LGBT population in the country, with 8.8% of residents identifying as homosexual, and the metro area ranks 4th in the nation at 6.1%.[134]

Racial history

Portland's population has been and remains predominantly white as a result of historic trends. In 2009, Portland had the fifth-highest percentage of white residents among the 40 largest U.S. metropolitan areas. A 2007 survey of the 40 largest cities in the U.S. concluded that Portland's urban core is the "whitest big city in the nation".[135] Some scholars have noted the Pacific Northwest as a whole is "one of the last Caucasian bastions of the United States".[136] While Portland's diversity was historically comparable to metro Seattle and Salt Lake City, those areas grew more diverse in the late 1990s and 2000s. Portland not only remains white, but migration to Portland is disproportionately white, at least partly because Portland is attractive to young college-educated Americans, a group which is overwhelmingly white.[135] [137]

The Oregon Territory banned African American immigration in 1849. In the 19th century, certain laws allowed the immigration of Chinese laborers but prohibited them from owning property or bringing their families.[135] [138] [139] The early 1920s saw the rapid growth of the Ku Klux Klan, which became very influential in Oregon politics, culminating in the election of Walter M. Pierce as governor.[138] [139] [140]

The largest influxes of minority populations occurred during World War II, as the African American population grew 10 times for wartime work.[135] After World War II, the Vanport flood in 1948 displaced many African Americans. As they resettled, redlining directed the displaced workers from the wartime settlement to neighboring Albina.[136] [139] [141] There and elsewhere in the Portland area, they experienced police hostility, lack of employment, and mortgage discrimination, leading to half the black population leaving after the war.[135] Widespread housing discrimination continues to affect the racial landscape today. A 2011 audit of landlords' renting practices by the Fair Housing Council of Oregon indicated that 64% of 50 leasing agents discriminated against Latino or black prospective tenants, compared with white prospective tenants. *The Oregonian* stated, "They were quoted higher rent and deposits, for example, or given additional fees, not offered applications or move-in specials, or shown inferior units."[142]

In the 1980s and 1990s, radical skinhead groups flourished in Portland.[139] During redevelopment of north Portland along the MAX Yellow Line, displacement of minorities occurred at a drastic rate. Out of 29 census tracts in north and northeast Portland, ten were majority nonwhite in 2000. By 2010, none of these tracts were majority nonwhite as gentrification drove the cost of living up.[143] However, Portland's African-American community still thrives in the north and northeast section of the city, mainly in the King Neighborhood.

Education

Portland is served by six public school districts and many private schools. Portland Public Schools is the largest school district. There are also many colleges and universities- the largest being Portland Community College, Portland State University, and Oregon Health & Science University. The city is also home to such private universities as the University of Portland, Reed College, and Lewis & Clark College.

Oregon Museum of Science and Industry

Oregon Museum of Science and Industry (OMSI), which includes many hands on activities for adults and children. OMSI consists of five main halls, most of which, consist of smaller laboratories: Earth Science Hall, Life Science Hall, Turbine Hall, Science Playground, and Featured Exhibit Hall. The Featured Exhibit Hall has a new exhibit every few months. The laboratories are Chemistry, Physics, Technology, Life, Paleontology, and Watershed. OMSI has many other unique attractions, such as the USS Blueback (SS-581), the OMNIMAX Dome Theater, and OMSI's Kendall Planetarium. The USS Blueback was the last non-nuclear fast attack submarine to join the US Navy and OMSI offers daily tours.[144] The OMNIMAX Dome Theater is a variant of the IMAX motion picture format, where the movie is projected onto a domed projection surface. The projection surface at OMSI's OMNIMAX Dome Theater is 6532 sq ft (606.8 m^2). The OMNIMAX Theater uses the largest frame in the motion picture industry and the frames are ten times the size of the standard 35mm film.[145] OMSI's Kendall Planetarium is the largest and most technologically advanced planetarium in the Pacific Northwest.[146] OMSI is located at 1945 SE Water Ave. OMSI is built right up next to the river and is also conveniently located near the entrance to the Springwater Corridor and Eastbank Esplanade pedestrian and bike trails.

Portland Art Museum

The Portland Art Museum owns the city's largest art collection and presents a variety of touring exhibitions each year and with the recent addition of the Modern and Contemporary Art wing it became one of the United States' twenty-five largest museums.

Oregon History Museum

The Oregon History Museum was founded in 1898. The Oregon History Museum has a variety of books, film, pictures, artifacts, and maps dating back throughout Oregon's history. The Oregon History Museum has one of the most extensive collections of state history materials in the USA.

The Portland Children's Museum

The Portland Children's Museum is a museum specifically geared for early childhood development. This museum has many topics, and many of their exhibits rotate, to keep the information fresh. The Portland Children's Museum also supports a small charter school for elementary children.

Crime

According to the Federal Bureau of Investigation's Uniform Crime Report in 2009, Portland ranked 53rd in violent crime out of the top 75 U.S. cities with a population greater than 250,000.[147] The murder rate in Portland over the last five years (2005–2009) has averaged 3.9 murders per 100,000 people per year, which is lower than the national average. For crimes other than murder, Portland is generally somewhat higher than the national average. According to the Portland Police, Killingsworth St., 82nd Ave., and the St. Johns Woods Apartments are the most dangerous areas of the city.[148] In October, 2009, the Forbes magazine rated Portland as the third safest city in America.[149] [150]

Sister cities

Portland has nine sister cities:[151]

- Ashkelon, Israel
- Kaohsiung, Republic of China (Taiwan)
- Ulsan, South Korea
- Mutare, Zimbabwe
- Guadalajara, Mexico
- Khabarovsk, Russia
- Sapporo, Japan
- Suzhou, People's Republic of China
- Bologna, Italy

Portland also has a "Friendship City" relationship with:

- Tallinn, Estonia[152]

Gallery

1897 topographic map of Portland shows streets, railroads, and significant differences in the Columbia Slough

The rose has a significant role in Portland's history, inspiring one of its nicknames.

MAX Light Rail and underground station at the Oregon Zoo

A view of the Willamette River from the Roof of the KOIN Center

Downtown Portland with Mt. Hood in the background

See also

- List of people from Portland, Oregon
- 1972 Portland-Vancouver Tornado
- List of hospitals in Portland, Oregon
- Keep Portland Weird

References

[1] McCall, William (August 19, 2003). "'Little Beirut' nickname has stuck". *The Oregonian*.

[2] "Elected Officials" (http://www.portlandonline.com/index.cfm?c=25999). City of Portland, Oregon. 2007. . Retrieved August 26, 2007.

[3] "American FactFinder" (http://factfinder.census.gov). United States Census Bureau. . Retrieved 2008-01-31.

[4] "US Board on Geographic Names" (http://geonames.usgs.gov). United States Geological Survey. 2007-10-25. . Retrieved 2008-01-31.

[5] http://www.portlandonline.com/

[6] "2010 Census profiles: Oregon cities alphabetically M-P" (http://www.pdx.edu/sites/www.pdx.edu.prc/files/media_assets/2010_PL94_cities_M-P.pdf) (PDF). Portland State University Population Research Center. . Retrieved October 2, 2011.

[7] Stevens, Suzanne (June 24, 2011). "Portland U.S.'s 23rd-largest metro area - Portland Business Journal" (http://www.bizjournals.com/portland/blog/2011/06/portland-uss-23rd-largest-metro-area.html?ed=2011-06-24&s=article_du&ana=e_du_pub). Bizjournals.com. . Retrieved 2011-09-19.

[8] "Find a County" (http://www.naco.org/Counties/Pages/FindACounty.aspx). National Association of Counties. . Retrieved 2011-06-07.

[9] "The "Smart Growth" Debate Continues" (http://www.innobriefs.com/editor/20030423smartgrowth.html). Urban Mobility Corporation. May/June 2003. . Retrieved November 7, 2006.

[10] Kate Sheppard (July 19, 2007). "15 Green Cities" (http://www.grist.org/article/cities3). Environmental News and Commentary. . Retrieved June 23, 2010.

[11] "Portland, Oregon: Green City of Roses | Frommers.com" (http://www.frommers.com/articles/1721.html). Frommers.com. . Retrieved October 20, 2008.

[12] "Portland – MSN Encarta" (http://encarta.msn.com/encyclopedia_761561539/portland.html). *Portland – MSN Encarta*. Encarta.msn.com. . Retrieved October 20, 2008.

[13] Marschner, Janice (2008). of portland before europeans indians&f=false *Oregon 1859: a spapshot in time* (http://books.google.com/?id=1-UW6M_gw48C&pg=PA187&dq=history+of+portland+before+europeans+indians#v=onepage&q=history). Timber Press. p. 187. ISBN 9780881928730. of portland before europeans indians&f=false.

[14] Orloff, Chet (2004). "Maintaining Eden: John Charles Olmsted and the Portland Park System". *Yearbook of the Association of Pacific Coast Geographers* **66**: 114–119. doi:10.1353/pcg.2004.0006.

[15] "Portland: The Town that was Almost Boston" (http://www.travelportland.com/media/history.html). Portland Oregon Visitors Association. . Retrieved November 18, 2006.

[16] Gibson, Campbell (June 1998). Population of the 100 Largest Cities and Other Urban Places in the United States: 1790 to 1990 (http://www.census.gov/population/www/documentation/twps0027.html). *U.S. Bureau of the Census – Population Division.*

[17] Loy, William G.; Stuart Allan, Aileen R. Buckley, James E. Meacham (2001). *Atlas of Oregon*. University of Oregon Press. pp. 32–33. ISBN 0-87114-101-9.

[18] "City keeps lively pulse." (Spencer Heinz, *The Oregonian*, January 23, 2001)

[19] City Flower (http://www.portlandonline.com/auditor/index.cfm?a=jbgc&c=cheid). *City of Portland Auditor's Office – City Recorder Division.*

[20] Stern, Henry (June 19, 2003). "Name comes up roses for P-town: City Council sees no thorns in picking 'City of Roses' as Portland's moniker". *The Oregonian.*

[21] "From Robin's Nest to Stumptown" (http://www.historicoregoncity.org/HOC/index.php?option=com_content&view=article&catid=70:oregon-trail-history&id=155:early-towns-and-cities&Itemid=98). End of the Oregon Trail Interpretive Center. . Retrieved January 23, 2012.

[22] "The Water" (http://web.archive.org/web/20061031090707/http://www.pdx.edu/water.html). Portland State University. Archived from the original (http://www.pdx.edu/water.html) on October 31, 2006. . Retrieved November 7, 2006.

[23] Baker, Nena (May 21, 1991). "R.I.P. FOR 'Rip City' Ruckus". *The Oregonian*: pp. A01

[24] Engel, Mary (May 30, 2010). "Achieving Beervana in Portland, Ore." (http://www.latimes.com/travel/la-tr-beer-20100530,0,767659.story). *Los Angeles Times*. . Retrieved May 30, 2010.

[25] Terry, Lynne (May 29, 2010). "Beervana gets shout out in L.A. Times" (http://www.oregonlive.com/portland/index.ssf/2010/05/beervana_gets_shout_out_in_la.html). *The Oregonian*. . Retrieved May 30, 2010.

[26] "Portland – Beer Town USA | DRAFT Magazine" (http://www.draftmag.com/beertowns/detail/portland). Draftmag.com. . Retrieved October 16, 2010.

[27] Hagestedt, Andre (April 7, 2009). "The Missing Oregon Coast: Waves After Dark" (http://www.beachconnection.net/news/missin040709_147.php). . Retrieved April 30, 2009. "I'm used to seeing that hint of dawn back in P-town, with my wretched habit of playing video games until 6 a.m"

[28] "Portland is new Soccer City, USA" (http://news.google.com/newspapers?id=II4QAAAAIBAJ&sjid=J-ADAAAAIBAJ&pg=5122,3289171&hl=en). *Eugene Register-Guard*. United Press International (Eugene, Oregon). August 13, 1975. . Retrieved June 22, 2010.

[29] Sandomir, Richard (November 6, 2008). "Seeking Help to Bring an M.L.S. Team to Portland" (http://www.nytimes.com/2008/11/07/sports/soccer/07franchise.html?ref=soccer). *The New York Times* (New York, New York). . Retrieved June 22, 2010.

[30] Sandomir, Richard (September 18, 2009). "Portland's ugly road to MLS status" (http://sportsillustrated.cnn.com/2009/soccer/09/17/portland.timbers/index.html). *Sports Illustrated*. . Retrieved June 22, 2010.

[31] Dure, Beau (August 26, 2009). "Portland Timbers show bark, bite as they prepare to join MLS" (http://www.usatoday.com/sports/soccer/2009-08-25-portland-timbers_N.htm). *USA Today* (McLean, Virginia). . Retrieved June 22, 2010.

[32] "US Gazetteer files: 2010, 2000, and 1990" (http://www.census.gov/geo/www/gazetteer/gazette.html). United States Census Bureau. 2011-02-12. . Retrieved 2011-04-23.

[33] "The Boring Lava Field, Portland, Oregon" (http://vulcan.wr.usgs.gov/Volcanoes/Oregon/BoringLavaField/description_boring_lava.html). USGS Cascades Volcano Observatory. . Retrieved November 7, 2006.

[34] "Mount Tabor Cinder Cone, Portland, Oregon" (http://vulcan.wr.usgs.gov/Volcanoes/Oregon/BoringLavaField/VisitVolcano/mount_tabor.html). USGS Cascades Volcano Observatory. . Retrieved April 20, 2007.

[35] Nokes, R. Gregory (December 4, 2000). "History, relived saved from St. Helens by a six-pack of Fresca". *The Oregonian*: p. 17.

[36] Kottek, M.; J. Grieser, C. Beck, B. Rudolf, and F. Rubel (2006). "World Map of the Köppen-Geiger climate classification updated" (http://koeppen-geiger.vu-wien.ac.at/pics/kottek_et_al_2006.gif). *Meteorol. Z.* **15** (3): 259–263. doi:10.1127/0941-2948/2006/0130. . Retrieved February 15, 2007.

[37] "GLOBAL ECOLOGICAL ZONING FOR THE GLOBAL FOREST RESOURCES ASSESSMENT 2000" (http://www.fao.org/docrep/006/ad652e/ad652e07.htm). Fao.org. . Retrieved October 16, 2010.

[38] "Climatography of the United States No. 20: 1971–2000" (http://cdo.ncdc.noaa.gov/climatenormals/clim20/or/356751.pdf). National Oceanic and Atmospheric Administration. . Retrieved September 15, 2010.

[39] "NOW Data-NOAA Online Weather Data" (http://www.wrh.noaa.gov/pqr/climate/pdx_clisummary.php). National Oceanic and Atmospheric Administration. . Retrieved July 30, 2009.

[40] "NowData - NOAA Online Weather Data" (http://www.nws.noaa.gov/climate/xmacis.php?wfo=pqr). National Oceanic and Atmospheric Administration. . Retrieved 2011-12-16.

[41] "Climatological Normals of Portland, OR" (http://www.weather.gov.hk/wxinfo/climat/world/eng/n_america/us/Portland_e.htm). Hong Kong Observatory. . Retrieved June 26, 2010.

[42] Houck, Mike. "Metropolitan Greenspaces: A Grassroots Perspective" (http://web.archive.org/web/20061003201150/http://www.audubonportland.org/conservation_advocacy/urbanconservation/metro_greenspaces). Audubon Society of Portland. Archived from the original (http://www.audubonportland.org/conservation_advocacy/urbanconservation/metro_greenspaces) on October 3, 2006. . Retrieved November 7, 2006.

[43] "Mount Tabor Park" (http://www.portlandonline.com/parks/finder/index.cfm?action=ViewPark&PropertyID=275). Portland Parks & Recreation. . Retrieved November 7, 2006.

[44] "America's 50 Greenest Cities" (http://www.popsci.com/environment/article/2008-02/americas-50-greenest-cities?page=1). . Retrieved June 10, 2010.

[45] "15 Green Cities" (http://www.grist.org/article/cities3/). . Retrieved June 10, 2010.

[46] "Imperial Sovereign Rose Court official site" (http://www.rosecourt.org/). . Retrieved December 8, 2008.

[47] "Lovecraft Film Festival Official Site" (http://www.hplfilmfestival.com/). . Retrieved November 25, 2007.

[48] "Kurt Cobain" (http://www.biography.com/articles/Kurt-Cobain-9542179?print). *Biography.com*. . Retrieved May 17, 2010.

[49] "Stumptown Comics Fest and Zine Library Group Unite" (http://www.graphicnovelreporter.com/content/stumptown-comics-fest-and-zine-library-group-unite-feature-stories). bookreporter.com. . Retrieved December 28, 2010.

[50] "2008 OBF biggest ever" (http://blog.oregonlive.com/thebeerhere/2008/07/2008_obf_biggest_ever_say_orga.html). *The Oregonian*. July 29, 2008. .

[51] Distefano, Anne Marie (July 8, 2005). "Brewers, beer lovers get many reasons to raise a glass" (http://www.portlandtribune.com/features/story.php?story_id=30717). *Portland Tribune*. .

[52] Merrill, Jessica (January 13, 2006). "In Oregon, It's a Brew Pub World" (http://travel.nytimes.com/2006/01/13/travel/escapes/13beer.html). *The New York Times*. . Retrieved September 29, 2010.

[53] "Oregon Experience: Beervana" (http://www.opb.org/programs/oregonexperiencearchive/beervana/player.php) (video). . Retrieved September 29, 2010.

[54] Ransom, Diana (September 16, 2011). "Why Portland's Beer Economy is 'Hoppy'" (http://www.entrepreneur.com/article/220319). Entrepreneur. . Retrieved 28 September 2011.

[55] Paul Toscano (September 8, 2010). "America's Best Cities for Happy Hour" (http://www.cnbc.com/id/39057498?slide=13). *CNBC*. . Retrieved September 29, 2010.

[56] "Portland: The center of the beer universe" (http://www.travelportland.com/media/mbmedkit/mb_beer.html). Portland Oregon Visitors Association. . Retrieved November 18, 2006.

[57] "Portland lifts a glass to its new name" (http://web.archive.org/web/20070213215818/http://koin.com/Global/story.asp?S=5932394). *KOIN 6 News*. January 12, 2006. Archived from the original (http://www.koin.com/Global/story.asp?S=5932394) on February 13, 2007. . Retrieved January 26, 2007.

[58] "TV : Food Network Awards : Food Network Awards Winners : Food Network" (http://www.foodnetwork.com/food/show_aw/text/0,3151,FOOD_28456_61089,00.html). Foodnetwork.com. . Retrieved October 6, 2008.

[59] Asimov, Eric (Published: September 26, 2007). "In Portland, a Golden Age of Dining and Drinking – New York Times" (http://www.nytimes.com/2007/09/26/dining/26port.html?pagewanted=3&_r=1). *The New York Times*. . Retrieved October 6, 2008.

[60] "America's Favorite Cities 2007 | Food/Dining | Food/Dining (Overall) | Travel + Leisure" (http://www.travelandleisure.com/afc/2007/category/6). Travelandleisure.com. . Retrieved October 6, 2008.

[61] "GoVeg.com // Features // North America's Most Vegetarian-Friendly Cities! // Portland, Oregon" (http://www.goveg.com/f-vegcities-portland.asp). Goveg.com. . Retrieved October 6, 2008.

[62] "World's Best Street Food" (http://travel.usnews.com/features/Worlds_Best_Street_Food/). U.S. News. .

[63] "World's Best Street Food" (http://articles.cnn.com/2010-07-19/travel/worlds.best.street.food_1_street-food-arepas-locals?_s=PM:TRAVEL). CNN Travel. November 3, 2010. .

[64] "A Few Favorite Portland Food Carts" (http://http://www.denverpost.com/travel/ci_15975669). DenverPost.com. . Retrieved September 14, 2010.

[65] "Portland Coffee Shops" (http://www.yelp.com/c/portland/coffee). Yelp.com. . Retrieved October 15, 2009.

[66] Strand, Oliver (September 16, 2009). "A Seductive Cup" (http://www.nytimes.com/2009/09/16/dining/reviews/16brief-001.html). *New York Times*. . Retrieved October 15, 2009.

[67] Neyer, Rob (August 21, 2003). "Though not perfect, Portland is a viable city for baseball" (http://sports.espn.go.com/mlb/columns/story?columnist=neyer_rob&id=1600284). ESPN. . Retrieved January 6, 2009. "Portland is the largest metropolitan area with just one major

professional sports team (the Trail Blazers)."
[68] "Oregon Sports Authority " Blog Archive Not every sport requires a ball" (http://www.oregonsports.org/not-every-sport-requires-a-ball/). Oregonsports.org. July 3, 2008. . Retrieved October 16, 2010.
[69] "Welcome to OCL" (http://www.oregoncricketleague.org). Oregon Cricket League. . Retrieved October 16, 2010.
[70] "Portland: Economy – Major Industries and Commercial Activity" (http://www.city-data.com/us-cities/The-West/Portland-Economy. html). . Retrieved June 4, 2008.
[71] "Quality of Living global city rankings 2009 – Mercer survey" (http://www.mercer.com/referencecontent.htm?idContent=1173105). Mercer. April 28, 2009. . Retrieved May 8, 2009.
[72] Kish, Matthew (January 20, 2011). "Notable Portland HQ departures" (http://www.bizjournals.com/portland/news/2011/01/20/ other-hq-companies-that-have-left-town.html?s=print). Portland Business Journal. . Retrieved January 25, 2011.
[73] "Metro: Urban growth boundary" (http://www.metro-region.org/index.cfm/go/by.web/id/277). . Retrieved June 4, 2008.
[74] Law, Steve (May 29, 2008). "Metro takes long view of growth" (http://www.portlandtribune.com/news/story. php?story_id=121200846357363500). Portland Tribune. . Retrieved June 4, 2008.
[75] "Portland – SkyscraperPage" (http://skyscraperpage.com/cities/?cityID=29). . Retrieved June 4, 2008.
[76] "OLMIS – Portland Metro Area: A Look at Recent Job Growth" (http://www.qualityinfo.org/olmisj/ArticleReader?itemid=00005735). . Retrieved June 4, 2008.
[77] "Steel Industry" (http://www.history.pdx.edu/guildslake/industry/steel1.htm). . Retrieved June 4, 2008.
[78] "Next stop: Port of Portland" (http://learfieldcreative.typepad.com/brownfield/2009/01/next-stop-port-of-portland-.html). January 7, 2009. . Retrieved February 6, 2009.
[79] "Port of Portland's Statement of Need" (http://www.ccrh.org/comm/slough/primary/stmntofnd.htm). Center for Columbia River History. . Retrieved February 6, 2009.
[80] "White House press release: The Columbia River Channel Deepening Project, August 13, 2004" (http://georgewbush-whitehouse.archives. gov/news/releases/2004/08/20040813-1.html). Georgewbush-whitehouse.archives.gov. August 13, 2004. . Retrieved October 16, 2010.
[81] "Cascade General, Inc" (http://www.answers.com/topic/cascade-general-inc?cat=biz-fin). . Retrieved June 4, 2008.
[82] "Portfolio" (http://www.portofportland.com/PDFPOP/Portfolio_06_07.pdf) (PDF). . Retrieved June 4, 2008.
[83] "2011 City and Neighborhood Rankings" (http://www.walkscore.com/rankings/cities/). Walk Score. 2011. . Retrieved Aug 28, 2011.
[84] "American Community Survey 2006, Table S0802" (http://factfinder.census.gov/servlet/ STGeoSearchByListServlet?ds_name=ACS_2008_1YR_G00_). U.S. Census Bureau. .
[85] "The Portland Streetcar Loop Project" (http://www.portlandstreetcar.org/node/11). portlandstreetcar.org. . Retrieved October 27, 2011.
[86] "Portland-Milwaukie Light Rail Bridge to bring new options for transit, cyclists and pedestrians" (http://www.trimet.org/pdfs/pm/ Fact-sheets-timelines/PMLR-Bridge-Fact-Sheet-August2011.pdf). trimet.org. . Retrieved October 27, 2011.
[87] "Trimet website" (http://www.myTrimet.com). Mytrimet.com. . Retrieved October 16, 2010.
[88] "Capital Campaign" (http://www.orhf.org/capital-campaign). ORHF.org. . Retrieved 31 December 2011.
[89] Ashton, David F. (20 December 2011). ""Holiday Express" delights families, benefits new S.E. museum" (http://www.thebeenews.com/ news/story.php?story_id=132423175015250800). The Sellwood Bee. . Retrieved 31 December 2011.
[90] "Portland Parks & Recreation, ORHF Mark the Start of Construction for New Enginehouse & Rail Heritage Center" (http://www. portlandonline.com/parks/index.cfm?a=370755&c=40197). Portland Parks & Recreation. 21 October 2011. . Retrieved 31 December 2011.
[91] "League of American Bicyclists * Press Releases" (http://www.bikeleague.org/media/press/). Bikeleague.org. . Retrieved October 6, 2008.
[92] "11 Most Bike Friendly Cities in the World – Bicycle friendly cities" (http://www.virgin-vacations.com/11-most-bike-friendly-cities. aspx). *Virgin Vacations*. Virgin Airlines. . Retrieved June 18, 2009.
[93] Bicycle Commute Challenge (http://www.bikecommutechallenge.com/oregon) information
[94] 'Youth Magnet' Cities Hit Midlife Crisis (http://online.wsj.com/article/SB124242099361525009.html) The Wall Street Journal. Retrieved on June 14, 2009.
[95] "19: Portland's Skatepark Master Plan" (http://skateportland.org/?page_id=11). Skaters for Portland Skateparks. . Retrieved July 18, 2006.
[96] Dougherty, Conor (July 30, 2009). "Skateboarding Capital of the World" (http://online.wsj.com/article/ SB10001424052970204119704574238073660408040.html). *The Wall Street Journal*. . Retrieved July 31, 2009.
[97] Mary Judetz, " Portland: Largest U.S. city with openly gay mayor (http://seattletimes.nwsource.com/html/localnews/ 2008581877_portlandmayor02m.html)" (January 2, 2009). Associated Press
[98] "Oregon Measure 36 Results by County" (http://www.uselectionatlas.org/RESULTS/state.php?year=2004&off=60&elect=0& fips=41&f=0). Uselectionatlas.org. . Retrieved October 16, 2010.
[99] "FBI's Joint Terrorism Task Force" (http://www.aclu-or.org/content/fbis-joint-terrorism-task-force). ACLU Oregon. April 28, 2005. .
[100] Byron York (November 28, 2010). "Politically correct Portland rejected feds who saved city from terrorist attack" (http:// washingtonexaminer.com/blogs/beltway-confidential/2010/11/politically-correct-portland-rejected-feds-who-saved-city-terrori). Washington Examiner. .
[101] "How Houston gets along without zoning – BusinessWeek" (http://www.businessweek.com/the_thread/hotproperty/archives/2007/ 10/how_houston_get.html). *Bloomberg BusinessWeek*. . Retrieved October 20, 2008.
[102] Sherry Thomas, special for USATODAY.com (Posted October 30, 2003 12:20 pm). "Houston: A city without zoning" (http://www. usatoday.com/travel/destinations/cityguides/houston/2003-10-07-spotlight-zoning_x.htm). *USA Today*. . Retrieved October 20, 2008.

[103] Author: Michael Lewyn. "Zoning Without Zoning | Planetizen" (http://www.planetizen.com/node/109). Planetizen.com. . Retrieved October 20, 2008.
[104] Robert Reinhold (Published: August 17, 1986). "FOCUS: Houston; A Fresh Approach To Zoning – New York Times" (http://query.nytimes.com/gst/fullpage.html?res=9A0DEFDB103FF934A2575BC0A960948260). Query.nytimes.com. . Retrieved October 20, 2008.
[105] Schadewald, Bill (April 9, 2006). "'The only major U.S. city without zoning' – Houston Business Journal:" (http://houston.bizjournals.com/houston/stories/2006/04/10/editorial1.html). Houston.bizjournals.com. . Retrieved October 20, 2008.
[106] Statewide Planning Goals. (http://www.oregon.gov/LCD/goals.shtml) Oregon Department of Land Conservation and Development. Retrieved December 23, 2007.
[107] "Comprehensive Land Use Planning Coordination" (http://www.leg.state.or.us/ors/197.html). Legislative Counsel Committee of the Oregon Legislative Assembly. . Retrieved November 7, 2006.
[108] Law, Steve (October 28, 2009). "Council adopts aggressive Climate Action Plan" (http://www.portlandtribune.com/news/story.php?story_id=125678469206117000). The Portland Tribune. . Retrieved August 1, 2010.
[109] "Grist 15 Green Cities" (http://www.grist.org/news/maindish/2007/07/19/cities/index.html). *Grist Magazine Online*. . Retrieved January 2, 2007.
[110] Sienstra (March 24, 2010). "Top 10 greenest cities: Portland makes the cut" (http://www.oregonlive.com/idahosportugal/index.ssf/2010/03/portland_makes_list_of_top_10.html). *The Oregonian*. . Retrieved March 24, 2010.
[111] Kipen, Nicki (March 24, 2010). "The Top 10 Greenest Cities" (http://www.move.com/home-finance/real-estate/general/top-greenest-cities-in-us.aspx). *Move.com*. . Retrieved March 24, 2010.
[112] Busse, Phil (November 7, 2002). "Cover Yourself!" (http://www.portlandmercury.com/portland/Content?oid=27886&category=22101). The Portland Mercury. . Retrieved February 1, 2007.
[113] Moore, Adam S.; Beck, Byron (November 8, 2004). "Bump and Grind" (http://www.wweek.com/story.php?story=6093). Willamette Week. . Retrieved February 1, 2007.
[114] Susan Donaldson James (October 22, 2008). "Strip Club Teases Small Oregon City—In National Capital of Stripping, Residents Say Free Speech Has Gone Too Far" (http://abcnews.go.com/TheLaw/Story?id=6088041&page=1). ABC News. . Retrieved December 8, 2008.
[115] "Judge: Salem lap dances protected by constitution" (http://www.katu.com/news/local/8263157.html). *KATU News*. Associated Press. June 30, 2007. . Retrieved December 8, 2008.
[116] "Is Portland 'Pornland?' Nightline highlights city sex trade" (http://www.katu.com/news/103648264.html). September 23, 2010. . Retrieved March 29, 2011.
[117] Rather, Dan (May 18, 2010). "Dan Rather: Pornland, Oregon: Child Prostitution in Portland" (http://www.huffingtonpost.com/dan-rather/pornland-oregon-child-pro_b_580035.html). *The Huffington Post*. . Retrieved March 29, 2011.
[118] Palahniuk, Chuck (2003). *Fugitives & refugees: a walk in Portland, Oregon*. New York: Crown Journeys. p. 101. ISBN 1-4000-4783-8.
[119] "Judge: riding in the buff is 'tradition,' man cleared" (http://www.katu.com/news/local/34445764.html). Associated Press. KATU. November 21, 2008. . Retrieved December 8, 2008.
[120] magnifiquem (June 12, 2009). "BUTTCRACKS AND BICYCLES: the Portland naked bike ride 2009!" (http://www.youtube.com/watch?v=i5wSKi5Bchk). YouTube. . Retrieved June 22, 2009.
[121] "Cyclists bare all in naked ride through Portland" (http://www.katu.com/news/local/48028862.html). *KATU*. June 14, 2009. . Retrieved June 22, 2009.
[122] Jonathan Maus, BikePortland (June 15, 2009). "Portland Naked Bike Ride: 5000 People" (http://pdxpipeline.wordpress.com/2009/06/15/portland-naked-bike-ride-5000-people-pictures-story/). *PDX Pipeline*. . Retrieved June 22, 2009.
[123] Jonathan Maus (June 14, 2009). "An estimated 5,000 take part in Portland's Naked Bike Ride" (http://bikeportland.org/2009/06/14/world-naked-bike-ride-was-it-good-for-you/). *Bike Portland*. . Retrieved June 22, 2009.
[124] Jonathan Maus (June 20, 2010). "An estimated 13,000 take part in Portland's Naked Bike Ride" (http://bikeportland.org/2010/06/20/an-estimated-13000-take-part-in-portlands-naked-bike-ride-35332). *Bike Portland*. . Retrieved November 29, 2010.
[125] "Oregon Court: Racist, insulting speech is protected" (http://www.seattlepi.com/local/375034_racist15.html). *Seattle Post-Intelligencer*. Associated Press. August 14, 2008. . Retrieved December 8, 2008.
[126] US Census Bureau State & County Quick Facts http://quickfacts.census.gov/qfd/states/41/4159000.html
[127] "State & County QuickFacts" (http://quickfacts.census.gov/qfd/states/41/4159000.html). U.S. Census Bureau. . Retrieved November 7, 2006.
[128] "Oregon – Place Population Estimates" (http://factfinder.census.gov/servlet/GCTTable?_bm=y&-geo_id=04000US41&-_box_head_nbr=GCT-T1-R&-ds_name=PEP_2009_EST&-_lang=en&-format=ST-9S&-_sse=on). U.S. Census Bureau. . Retrieved July 9, 2010.
[129] Egan, Timothy (March 24, 2005). "Vibrant Cities Find One Thing Missing: Children" (http://www.nytimes.com/2005/03/24/national/24childless.html?ei=5090&en=cbfa254535a51a5f&ex=1269320400&partner=rssuserland&pagewanted=all&position=). *The New York Times*. .
[130] MacColl, E. Kimbark (1979) [1979]. *The Growth of a City: Power and Politics in Portland, Oregon 1915–1950*. Portland, Oregon: The Georgian Press. ISBN 0-9603408-1-5.
[131] Management Information Services (2002). "Abernethy Elementary School: Recent Enrollment Trends, 1995–96 through 2002–03" (http://www.pps.k12.or.us/depts/mis/enroll/current/EnrollxSchl95-03.pdf). *Portland Public Schools*. . Retrieved September 1, 2010.

[132] "Vietnamese population by region: top metropolitan areas" (http://www.vatv.org/VAP.html). Vietnamese American Population. Retrieved January 7, 2011.
[133] "Pacific Islander" (http://www.pps.k12.or.us/files/esl/Pacific_Islander_Profile.pdf) (PDF), February 12, 2009. Retrieved January 5, 2011.
[134] Gary J. Gates Same-sex Couples and the Gay, Lesbian, Bisexual Population: New Estimates from the American Community Survey (http://www.law.ucla.edu/williamsinstitute/publications/SameSexCouplesandGLBpopACS.pdf)PDF (2.07 MB). The Williams Institute on Sexual Orientation Law and Public Policy, UCLA School of Law, October 2006. Retrieved April 20, 2007.
[135] Hammond, Betsy (September 30, 2009). "In a changing world, Portland remains overwhelmingly white" (http://www.oregonlive.com/news/index.ssf/2009/01/in_a_changing_world_portland_r.html). *The Oregonian*. . Retrieved March 11, 2011.
[136] Wilson III, Ernest J.; Wilson, Ernest J. (2004). *Diversity and US Foreign Policy: A Reader*. New York: Routledge. p. 55. ISBN 0-415-92884-2.
[137] Templeton, Amelia. "History Hinders Diversification Of Portland, Oregon : NPR" (http://www.npr.org/2011/02/16/133748144/history-hinders-diversification-of-portland-ore). *NPR*. . Retrieved March 11, 2011.
[138] Dresbeck, Rachel. *Insiders' Guide to Portland, Oregon* (7th ed.). p. 36. ISBN 9780762764754.
[139] Frazier, John W.; Tettey-Fio, Eugene L.. *Race, Ethnicity, and Place in a Changing America*. Global Academic Publishing. ISBN 1-58684-264-1.
[140] Levitas, Daniel (2002). *The terrorist next door: the militia movement and the radical right*. New York: Thomas Dunne Books/St. Martin's Press. ISBN 0-312-29105-1.
[141] Foster, Laura O.. *Portland Hill Walks: Twenty Explorations in Parks and Neighborhoods*. Timber Press, Incorporated. p. 239. ISBN 0-88192-692-2.
[142] Hannah-Jones, Nikole (May 9, 2011). "Portland housing audit finds discrimination in 64 percent of tests; city has yet to act against landlords" (http://www.oregonlive.com/portland/index.ssf/2011/05/a_portland_housing_audit_finds.html). *The Oregonian*. . Retrieved May 11, 2011.
[143] Hannah-Jones, Nikole (April 30, 2011). "Lessons learned? What Portland leaders did – and didn't do – as people of color were forced to the fringes" (http://www.oregonlive.com/pacific-northwest-news/index.ssf/2011/04/lessons_learned_what_portland_leaders_did_--_and_didnt_do_--_as_people_of_color_were_forced_to_the_f_1.html). *The Oregonian*. . Retrieved May 3, 2011.
[144] "USS Blueback: The Real Thing" (http://www.omsi.edu/submarine). . Retrieved July 22, 2010.
[145] "OMNIMAX Dome Theater" (http://www.omsi.edu/visit/omnimax). . Retrieved July 22, 2010.
[146] "OMSI Kendall Planetarium" (http://www.omsi.edu/planetarium). . Retrieved July 22, 2010.
[147] "Crime in the United States by Metropolitan Statistical Area, 2009 (Table 6)" (http://www.fbi.gov/about-us/cjis/ucr/crime-in-the-u.s/2009). FBI. . Retrieved October 12, 2010.
[148] Angela Valdez (July 5, 2006). "Night & Day – A photo essay of the most dangerous 'hood in Portland" (http://wweek.com/editorial/3235/7741/). *Willamette Week*. . Retrieved April 2, 2011.
[149] "America's Safest Cities for Real Estate" (http://www.forbes.com/2009/10/26/safest-cities-ten-lifestyle-real-estate-metros-msa.html). *Forbes*. October 26, 2009. . Retrieved October 16, 2010.
[150] "Crime Statistics" (http://www.cityrating.com/citycrime.asp?city=Portland&state=OR). *City Rating*. 2003. . Retrieved April 24, 2010.
[151] "About Portland's Sister Cities" (http://portlandonline.com/mayor/index.cfm?c=49918). Office of Mayor Sam Adams. . Retrieved March 18, 2009.
[152] "About the Sister City Program" (http://portlandonline.com/mayor/index.cfm?c=49951&a=235160). Portlandonline.com. March 13, 2009. . Retrieved October 16, 2010.

Further reading

- Abbott, C. (2001). *Greater Portland: Urban Life and Landscape in the Pacific Northwest*. Philadelphia: University of Pennsylvania Press. ISBN 0812217799.
- Abbott, Carl. *Portland in Three Centuries: The Place and the People* (Oregon State University Press; 2011) 192 pages; scholarly history
- Ozawa, C., ed (2004). *The Portland Edge: Challenges and Successes in Growing Communities*. Washington: Island Press. ISBN 1559636955.
- Palahniuk, Chuck (2003). *Fugitives and Refugees: A Walk in Portland, Oregon*. Crown. ISBN 1400047838.
- Holbrook, Stewart (1952). *The Far Corner*. Comstock Editions. ISBN 0891740430.
- MacColl, E. Kimbark (1976). *The Shaping of a City: Business and Politics in Portland, Oregon 1885 to 1915*. Portland: Georgian Press. OCLC 2645815.
- MacColl, E. Kimbark (1979). *The Growth of a City: Power and Politics in Portland, Oregon 1915 to 1950*. Portland: Georgian Press. ISBN 0960340815.

- Lansing, Jewel (2003). *Portland: People, Politics, and Power, 1851–2001*. Corvallis: Oregon State University Press. ISBN 0870715593.
- MacGibbon, Elma (1904). *Leaves of knowledge* (http://www.secstate.wa.gov/history/publications_detail.aspx?p=63). Shaw & Borden Co. Elma MacGibbons reminiscences of her travels in the United States starting in 1898, which were mainly in Oregon and Washington. Includes chapter "Portland, the western hub."
- O'Toole, Randal (July 9, 2007). "Debunking Portland: The City That Doesn't Work" (http://www.cato.org/pubs/pas/pa-596.pdf). *Policy Analysis* (Cato Institute) **596**.

External links

- City of Portland, Oregon (http://www.portlandonline.com/)
  - Portland Maps (http://www.portlandmaps.com/) (lot-level GIS)
  - Portland CrimeMapper (http://www.gis.ci.portland.or.us/maps/police/index.cfm)
- Portland Business Alliance – Portland Chamber of Commerce (http://www.portlandalliance.com/)
- Portland, Oregon (http://www.dmoz.org/Regional/North_America/United_States/Oregon/Localities/P/Portland//) at the Open Directory Project

Portland websites that are also wikis

- PortlandWiki (http://portlandwiki.org/PortlandWiki) is Portland, Oregon's civic wiki.
- WikiWikiWeb installed by Howard Cunningham from Beaverton. Since Ward invented the concept of a wiki wiki web, this is the very first wiki in existence.
- Portland (Oregon) travel guide from Wikitravel

Mercy Corps

Founder(s)	Ellsworth Culver, Dan O'Neill
Type	501(c)(3)
Founded	1979
Location	Portland, Oregon
Key people	Neal Keny-Guyer, Dan O'Neill Craig Redmond
Area served	Global
Motto	*Be the change.*
Website	www.mercycorps.org [1]

Mercy Corps is a global aid agency engaged in transitional environments that have experienced some sort of shock: natural disaster, economic collapse, or conflict. People working for it move as quickly as possible from bringing in food and supplies to enabling people to rebuild their economy with community-driven and market-led programs. To lay the groundwork for longer-term recovery, Mercy Corps focuses on connecting to both government and business for the changes they would like to see. "We actually focus on access to financial services as the critical element for helping to move people out of poverty", Nancy Lindborg, Mercy Corps President.[2]

Mercy Corps serves an area for extended periods of time to foster local entrepreneurship, rebuild social capital, and stimulate markets through "cash for work" programs and a variety of lending models. Mercy Corps, in the last 14 years, has founded 12 different finance institutions.[2] Since 1979, Mercy Corps has provided more than US$1.95 billion in assistance to people in 107 nations.[3] Supported by headquarters offices in North America and Europe, the agency's unified global programs employ 3,700 staff worldwide and reach nearly 16.7 million people in more than 40 countries.[3]

History

The organization was founded in 1979 as Save the Refugees Fund, a task-force organized by Dan O'Neill in response to the plight of Cambodian refugees fleeing the famine, war and genocide of the Killing Fields. By 1982, the organization had expanded its work to other countries, was joined by Ellsworth Culver (Mercy Corps co-founder), and was renamed Mercy Corps International to reflect its broader mission. After a shift from simply providing relief assistance to focusing on long-term solutions to hunger and poverty, Mercy Corps' first development project began in Honduras in 1982.

As of 2009, Mercy Corps' global headquarters are now located in the Old Town neighborhood of Portland, Oregon, at 45 SW Ankeny St. 97204. Opened October 9, 2009, the building houses the agency's headquarters offices, The Mercy Corps Action Center (a companion center to The Action Center to End World Hunger opened, October 2009 in Manhattan), Mercy Corps Northwest and The Lemelson Foundation.[4]

Mercy Corps also has a European Headquarters at 40 Sciennes, Edinburgh, Scotland EH9 1NJ. In 1996 Mercy Corps merged with Scottish European Aid and formed an organisation registered as Mercy Corps Scotland in 2000.

Activities

Mercy Corps Northwest

Mercy Corps' headquarters building in Portland, Oregon

The U.S. economic development office of Mercy Corps assists low-income individuals in Oregon and Washington who have difficulty receiving loans due to credit history, lack of collateral, or unfortunate circumstances. The loan program provides financing to entrepreneurs with solid business proposals and assists them to craft business plans and establish a matched savings account. Mercy Corps Northwest acts as a small business incubator and mentors new business owners through the start-up phase. MCNW also lends to existing businesses that create jobs for low-income people.[5]

Fostering microfinance

Partnering with banks and founding banking institutions, Mercy Corps facilitates microfinance around the world.[6]

Mercy Corps bank affiliates and area of operation:

- XacBank, Mongolia
- Partner (Partner Microcredit Organization), Bosnia and Herzegovina
- IMON International, Tajikistan
- Kompanion Financial Group LLC, Kyrgyzstan
- ACF (Asian Credit Fund), Kazakhstan
- AFS (Ariana Financial Services Group), Kabul
- AfK (Agency for Finance), Kosovo
- CFPA (Chinese Foundation for Poverty Alleviation), China
- MICRA (Microfinance Innovation Centre for Resources and Alternatives), Indonesia
- MAXIS (Maximizing Financial Access and Innovation at Scale) Indonesia and the Philippines
- Bank Andara, Indonesia
- PATRA (Poverty Alleviation in the Tumen River Area), China
- Borshud, Tajikistan
- CHAM (Community Health and Microcredit), Guatemala

Guatemala

Operations began in 2001 with a mother and child program at a health center. Mercy Corps expanded its work to mediating land conflicts. Nuevo Amanacer is one of several communities where Mercy Corps works to resolve disagreements over land and connect small-scale farmers to retail markets. The Inclusive Market Alliance for Rural Entrepreneurs was established by Mercy Corps, USAID, Walmart and Fundacion AGIL. The Alliance assists farmers to enlarge from subsistence farming to demand-driven production to supply major retailers like Walmart in Central America.[7] The Guatemala program operates out of Cobán, Alta Verapaz, with three sub-offices in Tucurú, Alta Verapaz, Nebaj, Quiche, and Guatemala City.[8]

Democratic Republic of Congo

To reduce violence against Congolese women vulnerable to abuse and rape when they go out collecting firewood and to alleviate deforestation of the Congo River basin forest, Mercy Corps introduced a fuel-efficient stove that requires less firewood. The stoves made from local clay, sand, and brick consume less than half the wood of traditional cooking fires. Mercy Corps teaches women how to make briquettes from manure and refuse which burn cleaner than charcoal. Local production of stoves and briquettes provides a way for women to make additional income and reduces the number of trips outside of the relative safety of the village. Reduction of carbon emissions through the use of the high-efficiency stoves has generated $160,000 worth of credits in the carbon market. Mercy Corps reinvests the proceeds in the education of Congolese women.[9]

North Korea

In response to the food crisis of the mid-nineties and at the request of the North Koreans, Mercy Corps began work in 1996. Mercy Corps has moved from food assistance to working on long-term agricultural and economic solutions: apple trees and fish farms and cross-cultural exchange. The effort involves aid workers living in Pyongyang, and in Chagang and North Pyongan provinces to visit families, monitor distribution and assess impact.[10]

Burma

After Cyclone Nargis in 2008, Mercy Corps arranged the shipment of emergency aid with the help of donations from DHL[11] and Chevron Corporation.[12] Government resistance to foreign aid kept supplies and relief workers confined to Rangoon.[13] Mercy Corps affiliated itself with Merlin, a British medical aid group which had a memorandum of understanding with the government.[13] Global aid agencies in the region rely on Burmese staff who are more able to move around obstacles created by the authoritarian regime.[13] Mercy Corps implemented "cash for work" programs paying residents to clean up flood-damaged villages and replant rice paddies.[14]

Finances

The agency's efficiency has helped Mercy Corps to achieve Charity Navigator's "Four-Star" charity rating (out of four possible stars).[15]

For the fiscal year ending June 30, 2006, Mercy Corps' income was US$205 million. US$81 million (40%) was from government grants. US$63 million (31%) was from in-kind donations ("material aid") in the form of food, medical, linens, and other supplies and services. Mercy Corps' expenses were US$191 million, which was spent primarily on the program but includes salaries and other compensation for 3,200 paid staff. Mercy Corps' total assets in 2006 were over US$98 million.

Organization, mergers

Mercy Corps is a non-governmental organization (NGO). Neal Keny-Guyer is the current Chief Executive Officer (CEO).

Mercy Corps incorporated the Conflict Management Group founded by Roger Fisher in 2004.[16] In 2007, Mercy Corps incorporated NetAid to expand its ongoing efforts to engage youth.[17]

Mercy Corps was also one of the first members of the IT nonprofit NetHope.[18]

Awards and distinctions

Awards and distinctions include:

- The Sitara-i-Eisaar (Star of Sacrifice) award, presented by the President of Pakistan, General Pervez Musharraf, in recognition of Mercy Corps' relief efforts following the 2005 Kashmir earthquake.[19] ;
- The Friendship Medal, presented posthumously to Mercy Corps co-founder Ellsworth Culver, by North Korean Ambassador Han Song Ryol - the first time that medal was presented to an American.[20] ;
- The 2003 Kahlil Gibran Spirit of Humanity Award, presented by the Arab American Institute Foundation for long-term efforts to help families living in the Arab world.[21]
- In 2008 Mercy Corps got "Appreciating A Global Partnership in Rebuilding Aceh-Nias" and "Sharing the Lesson" from President of the Republic of Indonesia Susilo Bambang Yudhoyono

References

[1] http://www.mercycorps.org
[2] Intelligent Investing with Steve Forbes, Interview with Nancy Lindborg, http://www.forbes.com/2009/12/16/lindborg-mercy-corps-intelligent-investing-video.html
[3] "2008 annual report" (http://www.mercycorps.org/2008annual). Mercy Corps. . Retrieved 2009-10-15.
[4] Bead, Richard (May 27, 2009). "As Mercy Corps grows, overhead has more than doubled since 2000". *The Oregonian.*
[5] Portland Monthly Magazine / Arts & Entertainment / Arts & Entertainment Articles / Detail (http://www.portlandmonthlymag.com/arts-and-entertainment/articles/mercy-corps-1009/1/)
[6] Microfinance Partners | Mercy Corps (http://www.mercycorps.org.uk/topics/microfinance/11281)
[7] http://guatemala.usembassy.gov/uploads/w0/iG/.../IMAREReleaseEng.pdf
[8] International Development - Global Aid Agency Goes Local in Guatemala (http://www.devex.com/articles/global-aid-agency-goes-local-in-guatemala)
[9] Heim, Kristi (2009-12-24). "Stoves aim to curb violence against women and the environment" (http://seattletimes.nwsource.com/html/thebusinessofgiving/2010587906_environmental_destruction_yet.html). *The Seattle Times.* .
[10] Mercy Corps helps get aid to N. Korea - OregonLive.com (http://www.oregonlive.com/news/oregonian/index.ssf?/base/news/121487551548280.xml&coll=7)
[11] Mercy Corps relies on DHL to ship emergency supplies worth $1 million from Portland to Myanmar and China (http://www.newswit.com/enews/2008-06-16/0837-mercy-corps-relies-on-dhl-to-ship/)
[12] Goldman Sachs | Media - Corporate Donations Bolster Mercy Corps' Dual Emergency Response Efforts (http://www2.goldmansachs.com/our-firm/press/press-releases/archived/2008/mercy-corps-emergency-announcement.html)
[13] Burmese Bloggers w/o Borders: Firsthand tale of Burma relief frustrations (http://bbwob.blogspot.com/2008/06/firsthand-tale-of-burma-relief.html)
[14] Snubbed U.S. mercy ships abandon Burma mission | Mail Online (http://www.dailymail.co.uk/news/worldnews/article-1024161/Snubbed-U-S-mercy-ships-abandon-Burma-mission.html)
[15] "Charity Navigator Rating - Mercy Corps" (http://www.charitynavigator.org/index.cfm?bay=search.summary&orgid=4078). Charity Navigator. . Retrieved 2009-12-12.
[16] Robert Perito (2007). *Guide for Participants in Peace, Stability, and Relief Operations.* United States Institute of Peace Press. p. 143. ISBN 978-1-60127-000-9.
[17] "Mercy Corps and NetAid Join Forces" (http://www.mercycorps.org/11334). Mercy Corps. January 2007. .
[18] About Us (http://www.nethope.org/about/us/) Official NetHope website.
[19] "Musharraf confers awards on outstanding Americans for relief work in quake-stricken areas" (http://www.presidentofpakistan.gov.pk/PRPressReleaseDetail.aspx?nPRPressReleaseId=1903&nYear=2007&nMonth=9) (Press release). Office of the Press Secretary to the President. 2006-09-22. . Retrieved 2007-11-13.
[20] "Pyongyang awards medal to American aid worker" (http://www.guardian.co.uk/international/story/0,,1687142,00.html). *The Guardian.* Associated Press (London). 2006-01-16. . Retrieved 2007-11-12.
[21] Kader, Nino. "Arab Foundation Awards Recognize "Spirit of Humanity"" (http://www.america.gov/st/washfile-english/2003/April/20030424164750tiwomods0.3930475.html). U.S. Department of State. . Retrieved 2007-11-12.

External links

- Mercy Corps United States (http://www.mercycorps.org/)

Pacific Northwest

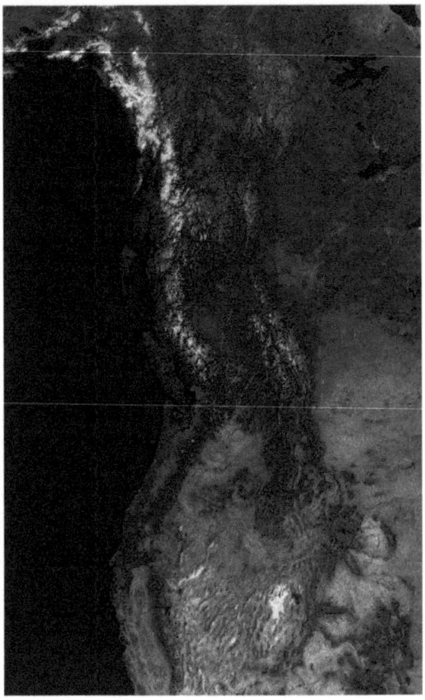

The Pacific Northwest from space

The **Pacific Northwest**, also known as **Cascadia** [1] [2] is a region in northwestern North America, bounded by the Pacific Ocean to the west and, loosely, by the Rocky Mountains on the east. Definitions of the region vary and there is no commonly agreed upon boundary, even among Pacific Northwesterners.[3] A common concept of the Pacific Northwest includes the U.S. states of Oregon, Washington and Idaho, as well as the Canadian province of British Columbia and territory of the Yukon.[4] This definition is often restricted further to include only the coastal areas west of the crest of the Cascade Mountains and Canadian Coast Mountains. Broader definitions of the region may even include the state of Alaska.[5] and may reach east to the Rocky Mountains.[4] Definitions based on the historic Oregon Country reach east to the Continental Divide, thus including nearly all of Idaho and parts of western Montana and western Wyoming. Sometimes the Pacific Northwest is defined as being the Northwestern United States, wholly in the United States. Often these definitions are made by government agencies whose scope is limited to the United States.[6] Some definitions include, in addition to Washington, Oregon, and British Columbia, Southeast Alaska, western Montana, the coast of northern California and a small part of northwestern Wyoming.[7] The term "Pacific Northwest" should not be confused with the Northwest Territory (also known as the Great Northwest, a historic term in the United States) or the Northwest Territories of Canada.

The term **Northwest Coast** is often used when referring only to the coastal regions. The term **Northwest Plateau** has been used to describe the inland regions, although they are commonly referred to as "the Interior" in British Columbia[8] and the Inland Empire in the United States.

The region's largest metropolitan areas are Seattle/Tacoma, Washington, with 3.3 million people;[9] Vancouver, British Columbia, with 2.3 million people;[10] and the Portland metropolitan area, with 2.2 million people.[11]

A key aspect of the Pacific Northwest is the US–Canada international border, which was established when the region was largely unsettled by non-indigenous peoples. The border along the 49th parallel and the Alaska Panhandle has had a powerful effect on the region. According to Canadian historian Ken Coates, the border has not merely influenced the Pacific Northwest—rather, "the region's history and character have been determined by the boundary".[4]

History

Indigenous peoples

The Pacific Northwest has been occupied by a diverse array of Indigenous American peoples for millennia, beginning with Paleo-Native Americans who explored and colonized the area roughly 15,000 years before Europeans arrived. The Pacific Coast is seen by a growing number of scholars as a major migration route for late Pleistocene peoples moving from northeast Asia into the Americas. Archaeological evidence for these earliest Native Americans is sketchy—in part because heavy glaciation, flooding, and post-glacial sea level rise have radically changed the landscape—but fluted Clovis-like points found in the region were probably left by Paleoindians at least 13,000 years ago. Even earlier evidence for human occupation dating back as much as 14,500 years ago is emerging from Paisley Caves in central Oregon.

With a history of human occupation spanning many millennia, and the incredible richness of Pacific Northwest fisheries (salmon, etc.), it is not surprising that the tribes who occupied the area historically were some of the most complex hunter-gatherer-fishers in history. They lived in large villages or towns, built plank houses and large canoes, and had sophisticated artistic and technological traditions. They were notable for being one of the few cultures to develop sedentary villages without practicing agriculture. In British Columbia and Southeast Alaska, for instance, maritime tribes like the Tlingit and Haida erected the large and elaborately carved totem poles that are iconic of Pacific Northwest artistic traditions. Throughout the area, thousands of descendants of these proud Pacific Northwest tribes still live and many of their cultural traditions continue to be practiced.

Initial European exploration

In 1579 the British captain and erstwhile privateer Francis Drake sailed up the west coast of North America perhaps as far as Oregon or even British Columbia before returning south to land and make ship repairs. At this landing site, probably near present-day San Francisco, Drake claimed the region for England, naming it New Albion.[5] [12] Juan de Fuca, a Greek captain in the employ of Spain, might have found the Strait of Juan de Fuca around 1592. The strait was named for him, but whether he discovered it or not has long been questioned.[13] During the early 1740s, Imperial Russia sent the Dane Vitus Bering to the region.[14] By the late 18th century and into the mid-19th century, Russian settlers had established several posts and communities on the northeast Pacific coast, eventually reaching as far south as Fort Ross, California. The Russian River was named after these settlements.

In 1774 the viceroy of New Spain sent Juan Pérez in the ship *Santiago* to the Pacific Northwest. Peréz made landfall on the Queen Charlotte Islands on July 18, 1774. The northernmost latitude he reached was 54°40′ N.[15] This was followed, in 1775, by another Spanish expedition, under the command of Bruno de Heceta and including Juan Peréz and Juan Francisco de la Bodega y Quadra as officers. On July 14, 1775 they landed on the Olympic Peninsula near the mouth of the Quinault River. Due to an outbreak of scurvy, Heceta returned to Mexico. On August 17, 1775 he sighted the mouth of the Columbia River but could not tell if it was a river or a major strait. His attempt to sail in failed due to overly strong currents. He named it *Bahia de la Asúnciõn*. While Heceta sailed south, Quadra continued north in the expedition's second ship, the *Sonora*. He reached 59° N, before turning back.[16]

In 1778 English mariner Captain James Cook visited Nootka Sound on Vancouver Island and also voyaged as far as Prince William Sound. In 1779 a third Spanish expedition, under the command of Ignacio de Artega in the ship *Princesa*, and with Quadra as captain of the ship *Favorite*, sailed from Mexico to the coast of Alaska, reaching 61° N. Two further Spanish expeditions, in 1788 and 1789, both under Esteban Jose Martínez and Gonzalo López de Haro, sailed to the Pacific Northwest. During the second expedition they met the American captain Robert Gray near Nootka Sound. Upon entering Nootka Sound, they found William Douglas and his ship the *Iphigenia*. There followed the Nootka Crisis, which was resolved by agreements known as the Nootka Convention. In 1790 the Spanish sent three ships to Nootka Sound, under the command of Francisco de Eliza. After establishing a base at Nootka, Eliza sent out several exploration parties. Salvador Fidalgo was sent north to the Alaska coast. Manuel

Quimper, with Gonzalo López de Haro as pilot, explored the Strait of Juan de Fuca, discovering the San Juan Islands and Admiralty Inlet in the process. Francisco de Eliza himself took the ship *San Carlos* into the Strait of Juan de Fuca. From a base at Port Discovery, his *pilotos* (masters) José María Narváez and Juan Carrasco explored the San Juan Islands, Haro Strait, Rosario Strait, and Bellingham Bay. In the process they discovered the Strait of Georgia and explored it as far north as Texada Island. The expedition returned to Nootka Sound by August 1791. Alessandro Malaspina, sailing for Spain, explored and mapped the coast from Yakutat Bay to Prince William Sound in 1791, then sailed to Nootka Sound. A scientific expedition in the manner of James Cook, Malaspina's scientists studied the Tlingit and Nuu-chah-nulth peoples before returning to Mexico. Another Spanish explorer, Jacinto Caamaño, sailed the ship *Aranzazu* to Nootka Sound in May 1792. There he met Quadra, who was in command of the Spanish settlement and Fort San Miguel. Quadra sent Caamaño north, to carefully explore the coast between Vancouver Island and Bucareli Bay, Alaska. Various Spanish maps, including Caamaño's, were given to George Vancouver in 1792, as the Spanish and British worked together to chart the complex coastline.[16]

From 1792 to 1794, George Vancouver charted the Pacific Northwest on behalf of Great Britain, including the Strait of Georgia, the bays and inlets of Puget Sound, and the Johnstone Strait–Queen Charlotte Strait and much of the rest of the British Columbia Coast and southeast Alaska shorelines.[15] For him the city of Vancouver and Vancouver Island are named, as well as Vancouver, Washington. From Mexico, Malaspina dispatched last Spanish exploration expedition in the Pacific Northwest, under Dionisio Alcalá Galiano and Cayentano Valdes aboard the schooners *Sutil* and *Mexicana*.[17] They met Vancouver in the Strait of Georgia on June 21, 1792. Vancouver had explored Puget Sound just previously. The Spanish explorers knew of Admiralty Inlet and the unexplored region to the south, but decided to sail north. They discovered and entered the Fraser River shortly before meeting Vancouver. After sharing maps and agreeing to cooperate, Galiano, Valdés, and Vancouver sailed north to Desolation Sound and the Discovery Islands, charting the coastline together. They passed through Johnstone Strait and returned to Nootka Sound. As a result, the Spanish explorers, who had set out from Nootka, became the first Europeans to circumnavigate Vancouver Island. Vancouver himself had entered the Strait of Juan de Fuca directly without going to Nootka first, so had not sailed completely around the island.[16]

In 1786 Jean-François de La Pérouse, representing France, sailed to the Queen Charlotte Islands after visiting Nootka Sound but any possible French claims to this region were lost when La Pérouse and his men and journals were lost in a shipwreck near Australia. Maritime fur trader Charles William Barkley also visited the area in the *Imperial Eagle*, a British ship falsely flying the flag of the Austrian Empire. American merchant sea-captain Robert Gray traded along the coast and discovered the mouth of the Columbia River.

Boundary disputes

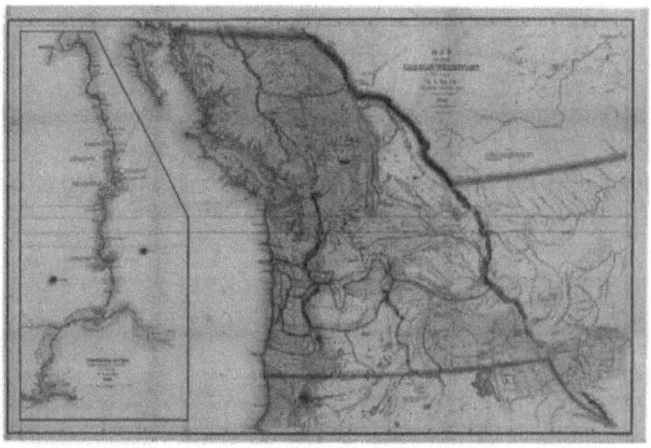

US Navy Admiral Charles Wilkes' 1841 Map of the Oregon Territory from "Narrative of the United States Exploring Expedition." Philadelphia: 1845

Initial formal claims to the region were asserted by Spain, based on the Treaty of Tordesillas which, in the Spanish Empire's interpretation, endowed that empire with the Pacific Ocean as a "Spanish lake". Russian Maritime Fur Trade activity, through the Russian-America Company, extending from the farther side of the Pacific to 'Russian America' prompted Spain to send expeditions north to assert Spanish ownership, while at the same time British claims were made and advanced by Captain James Cook and subsequent expeditions by George Vancouver. Potential French, Austrian and Portuguese claims were never advanced. As of the Nootka Conventions, the last in 1794, Spain gave up its exclusive *a priori* claims and agreed to share the region with the other Powers, giving up its garrison at Nootka Sound in the process.

The United States later established a claim following the exploration of the region by the Lewis and Clark Expedition, partly through the negotiation of former Spanish claims north of the Oregon-California boundary. From the 1810s until the 1840s, modern-day Washington, Oregon, Idaho, and western Montana, along with most of British Columbia, were part of what the United States called the Oregon Country and Britain called the Columbia District. This region was jointly claimed by the United States and Great Britain after the Treaty of 1818, which established a condominium of interests in the region in lieu of a settlement. In 1840 American Charles Wilkes explored in the area. John McLoughlin, Chief Factor of the Hudson's Bay Company, headquartered at Fort Vancouver, was the *de facto* local political authority for most of this time.

This arrangement ended as U.S. settlement grew and President James K. Polk was elected on a platform of calling for annexation of the entire Oregon Country and of Texas. After his election, supporters coined the famous slogan "Fifty-four Forty or Fight", referring to 54°40' north latitude—the northward limit of the region.[18] After a war scare with the United Kingdom, the Oregon boundary dispute was settled in the 1846 Oregon Treaty, partitioning the region along the 49th parallel and resolving most but not all of the border disputes (see Pig War).

The mainland territory north of the 49th parallel remained unincorporated until 1858, when a mass influx of Americans and others during the Fraser Canyon Gold Rush forced the hand of Colony of Vancouver Island's Governor James Douglas, who declared the mainland a Crown Colony, although official ratification of his unilateral action was several months in coming. The two colonies were amalgamated in 1866 to cut costs, and joined the Dominion of Canada in 1871. The U.S. portion became the Oregon Territory in 1848; it was later subdivided into territories that were eventually admitted as states, the first of these being Oregon itself in 1859. *See Washington Territory.*

American expansionist pressure on British Columbia persisted after the colony became a province of Canada, even though Americans living in the province did not harbor annexationist inclinations. The Fenian Brotherhood openly organized and drilled in Washington, particularly in the 1870s and the 1880s, though no cross-border attacks were experienced. During the Alaska Boundary Dispute, U.S. President Teddy Roosevelt threatened to invade and annex British Columbia if Britain would not yield on the question of the Yukon ports. In more recent times, during the so-called "Salmon War" of the 1990s, Washington Senator Slade Gorton called for the U.S. Navy to "force" the Inside Passage, even though it is not an official international waterway. Disputes between British Columbia and Alaska over the Dixon Entrance of the Hecate Strait between Prince Rupert and the Queen Charlotte Islands have not been resolved.[19]

Geology

Further information: Geology of the Pacific Northwest

The Northwest is still highly geologically active, with both active volcanoes and geologic faults.[20]

Active volcanoes in the region include Mount Garibaldi, Mount Baker, Mount Rainier, Mount Saint Helens, which erupted in 1980, Mount Hood, Mount Jefferson, and Mount Shasta.

Geography

The Pacific Northwest is a diverse geographic region, dominated by several mountain ranges, including the Coast Mountains, the Cascade Range, the Olympic Mountains, the Columbia Mountains, and the Rocky Mountains. The highest peak in the Pacific Northwest is Mount Rainier, in the Washington Cascades, at 14410 feet (4392 m). Immediately inland from the Cascade Range there is a broad plateau, narrowing progressively northwards, only a few miles wide in Canada, and also getting higher. In the US this region, semi-arid and often completely arid, is known as the Columbia Plateau, while in British Columbia it is the Interior Plateau, also called the Fraser Plateau. The Columbia Plateau was the scene of massive ice-age floods, as a consequence there are many coulees, canyons, and plateaus. Much of the plateau, especially in eastern Washington, is irrigated farmland. The Columbia River cuts a deep and wide gorge around the rim of the Columbia Plateau, and through the Cascade Range on its way to the Pacific Ocean. After the Mississippi, more water flows through the Columbia than any other river in the lower 48 states.

There are multiple definitions of the Pacific Northwest, none definitive. This map shows three possibilities. The shaded area shows the historical Oregon Country and the green line shows the Cascadia bioregion (a.k.a. Pacific Northwest bioregion), defined as "the watersheds of rivers that flow into the Pacific Ocean through North America's temperate rainforest zone".[21] A third definition is the labeled states, Washington, Oregon, and Idaho, and the province of British Columbia.

Because many areas have plentiful rainfall, the Pacific Northwest has some of North America's most lush and extensive forests, and at one time, the largest trees in the world. Coastal forests in some areas are classified as temperate rain forest.

The major cities of Vancouver, Portland, Seattle, and Tacoma all began as seaports supporting the logging, mining, and farming industries of the region, but have developed into major technological and industrial centers (such as the Silicon Forest), which benefit from their location on the Pacific Rim.

If defined as Oregon, Washington, and British Columbia, the Pacific Northwest has four US National Parks: Crater Lake in Oregon, and Olympic, Mount Rainier, and North Cascades in Washington. The Redwood National and State Parks are in California, just south of the Oregon border. Glacier Bay National Park is located in southeast Alaska, as is Wrangell–St. Elias National Park. Other outstanding natural features include the Oregon Coast, the Columbia River Gorge, the Columbia River, Mount St. Helens, and Hells Canyon. There are several Canadian National Parks in the Pacific Northwest, including Pacific Rim National Park on the west coast of Vancouver Island, Mount Revelstoke National Park and Glacier National Park in the Selkirk Range alongside Rogers Pass, Kootenay National Park and Yoho National Park on the British Columbia flank of the Rockies, Gwaii Haanas National Park Reserve in the Queen Charlotte Islands, and the Gulf Islands National Park Reserve in the Strait of Georgia. Although unprotected by national parks and only a handful of provincial parks, the south-central Coast Mountains in British Columbia contain the five largest mid-latitude icefields in the world.

Climate

The Pacific Northwest experiences a wide variety of climates. An Oceanic climate ("marine west coast climate") occurs in most coastal areas, typically between the ocean and high mountain ranges. An Alpine climate dominates in the high mountains. Semi-arid and Arid climates are found east of the higher mountains, especially in rainshadow areas. The Harney Basin of Oregon is an example of arid climate in the Pacific Northwest. Humid continental climates occur inland on windward sides, in places such as Revelstoke, British Columbia. A Subarctic climate can be found farther north, especially in Yukon and Alaska.

Under the Köppen climate classification, a cool summer version of the *dry-summer subtropical* (*Csb*) designation, typically referred to as "Mediterranean", is assigned to many areas of the Pacific Northwest as far north as southern Vancouver Island and the Gulf Islands, including cities such as Victoria, British Columbia, Seattle, and Portland.[22] These zones are not associated with a typical Mediterranean climate, and would be classified as Temperate Oceanic (*Cfb*), except dry-summer patterns typical to the Pacific Northwest meet Koeppen's minimum *Cs* thresholds. Other climate classification systems, such as Trewartha, place these areas firmly in the Oceanic zone (*Do*).[23]

Ecoregions

Much of the Pacific Northwest is forest. The Georgia Strait–Puget Sound basin is shared between British Columbia and Washington, and the Pacific temperate rain forests ecoregion, which is the largest of the world's temperate rain forest ecozones in the system created by the World Wildlife Fund, stretches along the coast from Alaska to California. The dryland area inland from the Cascade Range and Coast Mountains is very different from the terrain and climate of the Coast, and comprises the Columbia, Fraser and Thompson Plateaus and mountain ranges contained within them. The interior regions' climates are a northward extension of the Great Basin Desert which spans the Great Basin farther south, although by their northern reaches dryland and desert areas verge with boreal forest and various alpine flora regimes.

Population

Most of the population of the Pacific Northwest is concentrated in the Vancouver–Seattle–Portland corridor. This area is sometimes seen as a megacity (also known as a conurbation, an agglomeration, or a megalopolis). This "megacity" stretches along Interstate 5 in the states of Oregon and Washington and Hwy 99 in the province of British Columbia. As of 2004, the combined populations of the Greater Vancouver/Lower Mainland area, the Seattle metropolitan area and the Portland metropolitan area totaled almost nine million people.

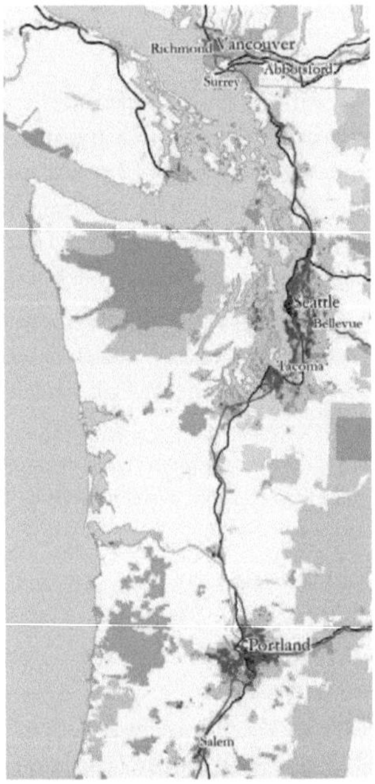

Map of "megacity", showing population density (shades of yellow/brown), highways (red), and major railways (black). Public land shown in shades of green.

Largest cities by population in the Pacific Northwest

City	State/Province	Population	Metropolitan Area
Vancouver	British Columbia	651,048[24]	2,404,911[24]
Seattle	Washington	608,660[25]	3,439,809[26]
Portland	Oregon	583,776[26]	2,226,009[26]
Surrey	British Columbia	473,238[27]	[28]
Spokane	Washington	208,916[25]	471,221[29]
Boise	Idaho	205,671[25]	616,561[26]
Tacoma	Washington	198,397[25]	[30]
Tri-Cities	Washington	181,756[29] [31]	253,340[26]
Vancouver	Washington	161,791[25]	[32]

Eugene	Oregon	156,185[25]	351,715[26]
Salem	Oregon	154,637[25]	390,738[26]
Abbotsford	British Columbia	123,864[33]	159,020[33]
Bellevue	Washington	122,363[25]	[30]
Kelowna	British Columbia	121,846[34]	187,234[34]
Gresham	Oregon	105,594[25]	[32]
Everett	Washington	103,019[25]	[30]
Hillsboro	Oregon	91,611[25]	[32]
Yakima	Washington	91,067[29]	243,231[29]
Beaverton	Oregon	89,803[25]	[32]
Nanaimo	British Columbia	86,961	
Victoria	British Columbia	84,031[35]	374,675[35]
Nampa	Idaho	81,557[25]	
Bellingham	Washington	80,885[25]	
Olympia	Washington	46,478[29]	252,264[26]

Politics

A major divide in political opinion separates the region's greatly more populated urban core and rural areas west of the mountains from its less populated rural areas to their east and (in B.C.) north.[36] The coastal areas—especially in the cities of Vancouver, Victoria, Bellingham, Seattle, Tacoma, Olympia, Portland, Corvallis, Eugene, and Ashland—are some of the most politically liberal parts of North America, consistently supporting left-wing political candidates and causes by significant majorities, while the Interior and North tend to be more conservative and consistently support right-wing candidates and causes. It should be noted that the religious right has far less influence throughout the region than elsewhere in the U.S., although it is very strong in the Fraser Valley, and also that certain areas of the BC Interior, particularly the West Kootenay and some areas of Vancouver Island and the BC Coast, have long histories of labour, environmental and social activism.

The urban core in addition to certain rural districts is known for supporting liberal political views, perceived as controversial in much of the rest of North America. Many jurisdictions have relatively liberal abortion laws, gender equality laws, legalized medical marijuana, and are supportive of LGBT rights, especially British Columbia, where gay marriage is legal, and Washington and Oregon, where same-sex civil unions are legal. Due to the urban core's size and voting impact, their counties and states as a whole have generally followed their leads, often to the disgruntlement of the more conservative rural areas. Oregon was the first U.S. state to legalize physician-assisted suicide, with the Death with Dignity Act of 1994. Washington State was the second when I-1000 passed in 2008. Colegio Cesar Chavez, the first fully accredited Hispanic college in the U.S., was founded in Mount Angel, Oregon in 1973. In 1986 King County, Washington, of which Seattle is a part, rebranded itself in honor of Martin Luther King.

These areas, especially around Puget Sound, have a long history of political radicalism. The radical labor organizers called Wobblies were particularly strong there in the mines, lumber camps and shipyards. A number of anarchist communes sprung up there in the early 20th century (see Charles Pierce LeWarne's Utopias on Puget Sound,

1885-1915 for an excellent overview of this popular yet forgotten movement). Seattle is one of a handful of major cities in North America in which the populace engaged in a general strike (in 1919) and was the first major American city to elect a woman mayor, Bertha Knight Landes (in 1926).[37] Socialist beliefs were once widespread (thanks in large part to the area's large numbers of Scandinavian immigrants) and the region has had a number of Socialist elected officials. So great was its influence that the U.S. Postmaster General, James Farley, jokingly toasted the "forty-seven states of the Union, and the Soviet of Washington", at a gala dinner in 1936 (although Farley denied ever saying it).[38]

The region also has a long history of starting cooperative and communal businesses and organizations, including Group Health,[39] REI, Puget Consumer's Co-ops and numerous granges and mutual aid societies. It also has a long history of publicly owned power and utilities, with many of the region's cities owning their own public utilities. In part as a result, the region enjoys the lowest electrical power rates on the continent. In British Columbia, credit unions are common and popular cooperatively owned financial institutions.

Economy

- Agriculture (fruit, potatoes, Tillamook cheese, dairy, wine, vegetables, wheat, Cascade hops, barley, hazelnuts)
- Aerospace (Boeing Commercial Airplane unit, Air Canada, Alaska Air, CHC Helicopter, Esterline)
- Diversified (Jim Pattison Group, Finning, Washington Marine Group)
- Entertainment industry (film and television, Lions Gate Entertainment, Lionsgate Studios, Lionsgate Television, Vancouver Film Studios, Bridge Studios)
- Finance and Banking (RBC, HSBC Bank Canada, Russell Investments, Umpqua Holdings Corporation)
- Forestry (Weyerhaeuser, Canfor, Tolko, Boise Cascade)
- Fishing and canning (salmon, halibut, herring, geoducks and other clams, crab, sea-urchin)
- High Technology and E-commerce (Microsoft, Microsoft Canada, Intel, F5 Networks, Nintendo of America, Nintendo of Canada, Tektronix, Amazon.com, Expedia, Ballard Power Systems, MacDonald Dettwiler, EA Canada, Cymax Stores, Micron Technology)
- Higher Education. There are several well-known colleges and universities in the Pacific Northwest.
  - British Columbia
    - List of colleges in British Columbia
    - List of universities in British Columbia
  - Washington
    - List of colleges and universities in Washington
  - Idaho
    - List of colleges and universities in Idaho
  - Oregon
    - List of colleges and universities in Oregon
  - Northern California
    - Humboldt State University
  - Montana
    - University of Montana
    - Montana State University System
- Hydroelectric power (Grand Coulee Dam, Bonneville Dam, Bridge River Power Project)
- Mass Retail (London Drugs, Costco, Blenz, Starbucks, Tullys, Nordstrom, Zumiez, Albertsons)
- Microbrewing (BridgePort, Deschutes, MacTarnahan's, Nelson, Ninkasi, Pyramid, Widmer Brothers, Yukon)
- Mining (Goldcorp, Cominco)

- Outdoor Tourism (Alpine Skiing, Snowboarding, Hiking, Kayaking, Rafting, Fishing, Mountain Biking, Water sports)
- Shoes & Apparel (Nike, Adidas North America, Columbia, R.E.I., Lululemon)
- Real estate marketing & realty development/construction.

Aluminum smelting was once an important part of the region's economy due to the abundance of once-cheap hydroelectric power and despite any bauxite reserves in the region. Hydroelectric power generated by the hydroelectric dams on the Columbia River powered at least ten aluminum smelters during the mid-20th century. By the end of World War II these smelters were producing over a third of the United States' aluminum. Production rose during the 1950s and 1960s, then declined. By the first decade of the 21st century the aluminum industry in the Pacific Northwest was essentially defunct.[40] The Alcan smelter at Kitimat continues in operation and is fed by the diversion of the Nechako River (a tributary of the Fraser) to a powerhouse on the coast at Kemano, near Kitimat.

The region as a whole, but especially the Seattle eastern suburbs along with Vancouver, is a hot-bed of high-tech business. It is also a leading "creative class" economic driver, with a thriving cultural sector, many knowledge workers and numerous international advertising, media and design firms.

Culture

While the dominant culture in the Pacific Northwest today is Anglo-American and Anglo-Canadian, there is significant Mexican and Chinese influence. 23% of Vancouver, B.C. is Chinese, and 50% does not speak English as a first language.[41] Parts of Oregon and Washington are bilingual in both English and Spanish, and Native American culture is strong throughout the Pacific Northwest. The hippie movement also began in California and the Pacific Northwest. There have been proposals for certain parts of the Pacific Northwest becoming its own country because of the shared ecoregion and culture.[42] [43] The most well known proposals are Ecotopia from the Nine Nations of North America and Cascadia. However, the region is strongly divided by the international border, and this division has grown more rather than less powerful over the 20th century.[44] In addition, although the metropolitan centers of Vancouver, Seattle, and Portland are bound into a kind of urbanized region, they originated and continue to thrive as east-west gateways, competing with each other, rather than north-south connectors.[44]

Environmentalism

Environmentalism is prominent throughout the region, especially west of the Cascades. Environmentally conscious services such as recycling and public transportation are widespread, most notably in the more populous areas. A recent statistical analysis ranked the 50 Greenest Cities in the United States, placing Portland, Oregon first, Eugene, Oregon fifth, and Seattle, Washington eighth.[45] The region as a whole is also known for its bicycle culture as an alternative form of transportation; Portland is considered the second most bicycle-friendly city in the world.[46] Portland is also the hub of American bicycle manufacturing; as a whole it generated over $68 million in revenue in 2007 alone.[46] Politically, the Pacific Northwest is actively involved in environmental efforts. The international organization Greenpeace was born in Vancouver in 1970 as part of a large public opposition movement in British Columbia to US nuclear weapons testing on Amchitka Island in the Aleutians. Liberal and Conservative Northwesterners, such as former U.S. Senator Slade Gorton (R-WA) and moderate Democrats like former Speaker of the House Tom Foley (D-WA), have been prominent in the development of conservative approaches to environmental protection. Seattle in particular is also home to a large number of publications and institutions concerned with the environment and sustainability, including both *Worldchanging* and *Grist.org*, the U.S.'s two largest online green magazines. The Pacific Northwest is also noted for a large number of gardening clubs, with Victoria having an annual flower count in February.

Music

The modern-era Pacific Northwest is known for indie music, especially grunge and alternative rock, as well as historically-strong folk music and world music traditions. Many are associated with the famous independent label Sub Pop. KEXP.org is a popular and nationally-noted Seattle-based public indie music radio station. Among the area's largest music festivals are the Merritt Mountain Music Festival, the Vancouver Folk Music Festival, the Sasquatch! Music Festival in George, Washington, Seattle's Bumbershoot, and Portland's Musicfest NW. Portland's Waterfront Blues Festival is the largest festival west of the Mississippi.

Local artists who became ground-breaking rock bands of their times include:

- Jimi Hendrix
- The Kingsmen
- Nirvana
- Soundgarden
- Presidents of The United States of America
- Heart
- Queensrÿche
- Built to Spill
- Death Cab for Cutie
- DOA
- Foo Fighters
- Agalloch
- Elliott Smith
- The Decemberists
- The Dandy Warhols
- Sleater-Kinney
- Modest Mouse
- Cherry Poppin Daddies
- Everclear
- Macklemore
- Pedro the Lion
- Alice in Chains
- the Subhumans
- Skinny Puppy
- Front Line Assembly
- Blitzen Trapper
- Pink Martini
- Postal Service
- Pearl Jam

Other artists worthy of note who hail from the Pacific Northwest are Bryan Adams, Nelly Furtado, Swollen Members, Neko Case, Sarah MacLachlan, Michael Bublé, Blue Scholars, Fleet Foxes, and Meredith Brooks.

Cuisine

Cuisine of the area include wild salmon, huckleberries, a wide variety of Asian cuisines, and locally-produced fruits, vegetables, and cheeses. Chinese, Japanese, Korean, Italian and Greek cuisines are prevalent throughout the Northwest (most especially in Vancouver), and reflect the historically strong presence of those communities in the restaurant industry there. Similarly eateries featuring Persian, Asian Fusion, and Indo-Canadian cuisines are common throughout in Greater Vancouver, as are ethnic specialty restaurants of all kinds. Ethnic staples ranging from frozen perogies to frozen dim sum are common in most supermarkets in these communities.

Locally-made craft beers and premium wines from various wine-growing area within the region are popular with drinkers and diners. Northern latitude and coastal breezes create a climate that attracts international recognition for its mostly family-owned and operated vineyards and wineries.

Portland is considered to be the microbrew capital of America,[47] and is home to the Widmer Brothers Brewery.

Cannabis use is relatively popular, especially around Vancouver BC, Bellingham, Seattle, Olympia, Spokane, Portland and Eugene. Several of these jurisdictions have made arrests for cannabis a low enforcement priority. Medical marijuana is legal in British Columbia,[48] Washington[49] and Oregon.[50]

Sports

Skiing, snowboarding, cycling, mountaineering, hiking, camping, hunting, fishing, boating, and water sports are popular outdoor activities. Vancouver, Seattle, Tacoma, Portland and Spokane are home to numerous professional sports teams, including the BC Lions, Vancouver Canucks, Vancouver Canadians, Vancouver Whitecaps FC, Seattle Mariners, Seattle Seahawks, Seattle Sounders FC, Seattle Storm, Washington Stealth, Everett Aquasox, Tacoma Rainiers, Portland Trail Blazers, Portland Timbers, Spokane Shock and Spokane Indians. The Pacific Northwest is known for being one of the most passionate hotbeds for soccer in the United States. The region's three Major League Soccer teams the Whitecaps, Sounders, and Timbers play to sold out crowds and compete annually for the Cascadia Cup.

CenturyLink Field, home of Seattle Seahawks and Sounders FC

Sports fans in Oregon are particularly passionate; followers of the Portland Trail Blazers basketball team refer to themselves as the *Sixth Man* and "Blazermania" is a term used to describe the extraordinary dedication fans have shown the team. In Seattle many fans are still upset over the move of the Seattle Sonics while supporters of the Seattle Seahawks football team are known officially as the 12th Man.[51] And the supporter groups, (namely the Emerald City Supporters, Timbers Army, and Southsiders) of the three MLS teams of the region are renowned for their passion and dedication to their teams.

In Washington and Oregon, many residents passionately follow college athletics. In Washington, the major NCAA Division I college athletic programs are the University of Washington Huskies and the Washington State Cougars. In Oregon the major programs are the University of Oregon Ducks and the Oregon State Beavers. All four of these programs are members of the Pacific-12 Conference and compete with each other in a variety of sports. These universities are all considered to be rivals of one another, particularly in college football. The most significant of these rivalries are the Oregon – Washington game, the Washington-Washington State game, a.k.a the Apple Cup (named so because of Washington's notoriety for apple production) and the Oregon-Oregon State game a.k.a the Civil War. As in professional sports, college fans in the Pacific Northwest are known for being particularly passionate about their teams. Both Husky Stadium (where the Washington Huskies play football) and Autzen Stadium (where the Oregon Ducks play football) have gained reputations for deafening noise, despite not being the largest of college football venues. Husky Stadium currently holds the record for the loudest crowd noise in NCAA history at 130 decibels,[52] while Autzen Stadium currently holds the record for the 4th at 127 decibels.[53]

Video games

Seattle is considered by *Digital Trends* magazine to be the top gaming city in America, a possible indicator of markedly higher rates of video game usage throughout the Pacific Northwest in general.[54] Major companies headquartered in the region include Microsoft (Seattle Metropolitan Area), Microsoft Canada (Vancouver), Nintendo of America (Seattle Metropolitan Area), Nintendo of Canada (Vancouver), and video games maker Electronic Arts (Vancouver), as well as VALVe (Seattle Metropolitan Area).

Demographics

In the US side of the region, Latinos make up a large portion of the agricultural labor force east of the Cascade Range, and are an increasing presence in the general labor force west of the Cascades. African Americans are less numerous in the Pacific Northwest, however the overall African American population has been growing in other smaller urban areas throughout the region, such as Spokane and Eugene.[55] African Americans tend to be concentrated in western urban areas such as Tacoma, south Seattle, and Portland. Nonetheless, blacks have a very large presence in Tacoma's Hilltop and South Tacoma neighborhoods, Seattle's Central District and Rainier Valley neighborhoods and in Portland's Northeast Quadrant. There are growing numbers of Africans in Vancouver BC as well as Jamaicans and blacks from the US. As of the first decade of the 21st century, many Asians were moving out and into middle class suburbs, though some would voice concern about preserving historical communities particularly in Vancouver. British Columbia has the largest Asian presence per-capita in North America, with 10% of the population being of Chinese ancestry and also large numbers of South Asians, Filipinos, and others. The Asian presence in the U.S. section of the Pacific Northwest is comparably smaller, with all Asian groups together comprising about 8% of Washington state's population, and smaller figures in Oregon and Idaho.

African-Americans have held the positions of Mayor in Seattle and Spokane; King County executive, while the state of Washington elected a Chinese American Governor during the 1990s, Gary Locke.

British Columbians of many ethnicities are prominent in all levels of politics and government, and the province has a number of "firsts" in Canadian political history, including the first non-white Premier, Ujjal Dosanjh (who is Indo-Canadian) and the first Asian Lieutenant-Governor, the Hon. David Lam. The current Lieutenant-Governor, Steven Point, is of aboriginal origin, being Stó:lō (the dominant type of Coast Salish in BC's Lower Mainland) from the Chilliwack area. The former leader of the opposition party, the NDP, is Carole James, who is of partial Métis origin. Colonial governor James Douglas was himself mulatto of Guyanese extraction and his wife was of Cree origin.

On LGBT representation in government, Oregon has been a national leader. At the time of his election to the office of Portland mayor in 2008, Sam Adams was the first openly gay individual to represent a city of Portland's size in the United States. In Silverton, Oregon, the same year, Stu Rasmussen was elected the first transgendered mayor in U.S. history. The first two LGBT state supreme court justices in the country both sit on the Oregon Supreme Court. Oregon Secretary of State Kate Brown, who is first in line to the governorship, is the highest-ranking openly bisexual politician in the U.S.

Language

The Pacific Northwest English accent is considered to be "very neutral" to most Americans. It does, however, possess the low back vowel merger, or the cot–caught merger. Pacific Northwest English is one of the closest living accents to conservative General American English. It lacks the Northern Cities Vowel Shift, and does not participate as strongly in the California Vowel Shift or the Canadian raising as do other regional accents. Because of its lack of any distinguishing vowel shift, the accent is very similar to and hard to distinguish from conservative speakers in other dialect regions especially the Northern Midlands, California, and the prairies.

Chinook Jargon was a pidgin or trade language established among the indigenous inhabitants of the region. After contact with Europeans, French, English and Cree words entered the language, and "eventually Chinook became the lingua franca for as many as 250,000 people along the Pacific Slope from Alaska to Oregon".[56] Chinook Jargon reached its height of usage in the 19th century though remained common in resource and wilderness areas, particularly but not exclusively by Native Americans and Canadian First Nations people, well into the 20th century. Today its influence is felt mostly in place names and a handful of localized slang terms, particularly the word skookum, which remains hallmark of people raised in the region.

Besides English and indigenous languages, Chinese has been common since the gold rushes of the mid-19th century, most particularly in British Columbia. Since the 1980s the Toishan, a Cantonese-based dialect which was predominant in the area, has been replaced by mainstream Cantonese and by Mandarin because of large-scale immigration from Asia. Punjabi is also common in Vancouver, which has very large Sikh communities.

Spirituality and religion

The Pacific Northwest has the lowest rate of church attendance in the United States and consistently reports the highest percentage of atheism;[57] [58] this is most pronounced on the part of the region west of the Cascades.[59] A recent study indicates that one quarter of those in Washington and Oregon believe in no religion.[60]

Religion plays a smaller part in Pacific Northwest politics than in the rest of the United States. The religious right has considerably less political influence than in other regions. Political conservatives in the Pacific Northwest tend to identify more strongly with free-market libertarian values than they do with more religious social conservatives.[61]

That said, three of the four major international charities in the region are religious in nature: Northwest Medical Teams International, World Concern, World Vision International, and Mercy Corps. This is part of a long tradition of activist religion. The Skid Road group, a shelter offering soup and sermons to the unemployed and recovering alcoholics, was launched in Vancouver, with the Salvation Army having deep roots in the Gastown district, dating back to the era of the construction of the Canadian Pacific Railway (1880s) and attained prominence in the same centers during the Klondike Gold Rush.

The region is also known as a magnet for a wide range of philosophical and spiritual belief systems. Eastern spiritual beliefs have been adopted by an unusually large number of people (by North American standards), and Tibetan Buddhism in particular has a strong local following.[62] The Northwest Tibetan Cultural Association, claimed to be the largest organization of its kind in the world, was founded in Portland in 1993.

The region is home to many unique Christian communities, ranging from the Doukhobors to the Mennonites. The Mennonite Central Committee Supportive Care Services is based in Abbotsford, BC.[63] Mennonite Central Committee and Mennonite Disaster Service enjoy a heavy rate of enlistment and donations from the strong Mennonite community in British Columbia's Fraser Valley. Also within the region there is a fairly strong representation of Orthodox churches (Greek, Russian, Serbian and others), as well as the Ukrainian Greek Catholic Church. Religious sees that are based in the Pacific Northwest include the Roman Catholic ecclesiastical provinces of Portland, Seattle, and Vancouver, Province 8 of the Episcopal Church in the United States of America, the Anglican Ecclesiastical Province of British Columbia and the Yukon, and the suffragan dioceses that make up those provinces.

Yogic teachings, Sufism, tribal and ancient beliefs and other philosophies are widely studied and appreciated in the region. The Lower Mainland of British Columbia has a very large Sikh community. Oregon has a considerable Quaker population. See Society of Friends. There has been major growth in Chinese Buddhist temples since the increase in immigration from East Asia in the 1980s, especially in Vancouver.

Also in Vancouver, there is a small Hindu population, a number of Parsee (Zoroastrians), and an emerging Muslim population from South Asia, the Middle East, Africa, the Balkans, Southeast Asia and elsewhere.

Some people in the area also embrace alternative religion, such as New Age spirituality and Neo-Paganism.[64]

- Before its closure in 2004, Mary Manin Morrissey's "megachurch" called Living Enrichment Center, located in Wilsonville, Oregon, was one of the biggest New Thought churches in the entire world, with a congregation estimated at between two thousand and five thousand members. Morrissey's "Life Keys" religious program was broadcast to several major networks around the U.S. West Coast.

- Neale Donald Walsch, author of *Conversations with God*, lives in Ashland, Oregon, where he runs a retreat center.
- Gangaji, an internationally recognized spiritual teacher and disciple of Poonjaji, lives in Ashland, Oregon.
- Established in more recent times, the training school of the immortal (according to the organization) being Ramtha is headquartered in Yelm, Washington.
- The followers of the Guru Rajneesh, the *sannyasins*, established a center for their beliefs and lifestyle near Antelope, Oregon, which included an ashram complex as well as, for a while, an attempted takeover of the local economy.
- The Emissaries of Divine Light are a notable presence in the region of 100 Mile House, British Columbia.
- More controversially, the commune run by Brother Twelve in the Gulf Islands of British Columbia early in the 20th century.
- Oregon's Willamette Valley has a large population of Russian Old Believers.[65]

See also

- Atlantic Northeast, another region shared between Canada and the US
- Cascadia (disambiguation)
- Climate change in Washington
- Megaregions of the United States
- Northwest Coast art
- Cascadia (independence movement)

Notes and references

Notes

[1] http://www.youra.com/cascadia/

[2] http://www.peopleofcascadia.com/

[3] Schwantes, Carlos A. (1 January 1996). *The Pacific Northwest: an interpretive history* (http://books.google.com/books?id=JImlIbueaXcC&pg=PA1). University of Nebraska Press. pp. 1–2. ISBN 9780803292284. . Retrieved 14 December 2010.

[4] Coates, Ken S. (2002). "1. Border Crossings" (http://books.google.com/books?id=4leTYpWOE5UC&pg=PA3). In Findlay, John M.; Coates, Ken S.. *Parallel Destinies: Canadian-American Relations West of the Rockies*. University of Washington Press. pp. 3–5. ISBN 9780295982526. . Retrieved 14 December 2010.

[5] Hayes, Derek (1999). *Historical atlas of the Pacific Northwest: maps of exploration and discovery : British Columbia, Washington, Oregon, Alaska, Yukon* (http://books.google.com/books?id=sl57oHrVXGoC&pg=PA8). Sasquatch Books. p. 8. ISBN 9781570612152. . Retrieved 14 December 2010.

[6] See, for example, this US [[United States Environmental Protection Agency|EPA (http://www.nbii.gov/portal/server.pt/gateway/PTARGS_0_2_25589_1921_7072_43/http;/cbi-lap7.cbi.cr.usgs.gov;7097/publishedcontent/publish/geographic/southwest/pnwin_regional_ecosystems_intro/regionalecosystems_pnwin_l2_resized.jpg)] map].

[7] Apostol, Dean; Sinclair, Marcia; Society for Ecological Restoration International (2006). *Restoring the Pacific Northwest: the art and science of ecological restoration in Cascadia* (http://books.google.com/books?id=CsGyhzFBjyAC&pg=PR24). Island Press. p. 24. ISBN 9781559630788. . Retrieved 14 December 2010.

[8] "Interior Savings Centre in Kamloops, British Columbia" (http://www.interiorsavingscentre.com/index.shtml). Interiorsavingscentre.com. . Retrieved 2010-08-18.

[9] "Best Places to Live in Seattle-Tacoma-Bellevue, Washington Metro Area (WA)" (http://www.bestplaces.net/Metro/Seattle-Tacoma-Bellevue-Washington.aspx). Bestplaces.net. . Retrieved 2010-08-18.

[10] "http://www40.statcan.gc.ca/l01/cst01/DEMO05A-eng.htm" (http://www.metrovancouver.org/REGION/TEACHERS/Pages/default.aspx). .

[11] "Doing Business In Portland, Oregon" (http://www.business.gov/states/oregon/local/portland.html). *Business.GOV*. .

[12] Quanchi, Max; Robson, John (2005). *Historical dictionary of the discovery and exploration of the Pacific islands* (http://books.google.com/books?id=q20QAJEDcScC&pg=PA207). Scarecrow Press. pp. 84–85, 207. ISBN 9780810853959. . Retrieved 12 July 2011.

[13] Hayes, Derek (1999). *Historical atlas of the Pacific Northwest: maps of exploration and discovery : British Columbia, Washington, Oregon, Alaska, Yukon* (http://books.google.com/books?id=sl57oHrVXGoC&pg=PA16). Sasquatch Books. p. 16. ISBN 9781570612152. . Retrieved 12 July 2011.

[14] "Vitus Bering Biography Summary" (http://www.bookrags.com/Vitus_Bering). BookRags.com. . Retrieved 2010-08-18.

[15] "Historical Timeline" (http://www.ubcic.bc.ca/Resources/timeline.htm). Ubcic.bc.ca. 2007-09-13. . Retrieved 2010-08-18.

[16] Hayes, Derek (1999). *Historical Atlas of the Pacific Northwest: Maps of exploration and Discovery*. Sasquatch Books. ISBN 1-57061-215-3.

[17] "Captain Alexandro Malaspina" (http://www.mala.bc.ca/www/discover/capt/capt.htm). Malaspina University-College. . Retrieved 2008-02-05.

[18] "Fifty-Four Forty or Fight" (http://geography.about.com/od/politicalgeography/a/5440orfight.htm). Geography.about.com. 2010-06-14. . Retrieved 2010-08-18.

[19] "The Alaska Boundary Dispute" (http://www.cwilson.com/pubs/energy/alaska.pdf) (PDF). . Retrieved 2010-08-18.

[20] "Volcano Seismicity in the Pacific Northwest" (http://www.pnsn.org/INFO_GENERAL/volcanoes.html). Pnsn.org. . Retrieved 2010-08-18.

[21] "Map of Cascadia" (http://www.sightline.org/maps/maps/cascadia_cs05m). Sightline.org. . Retrieved 2011-02-19.

[22] "HESS - Abstract - Updated world map of the Köppen-Geiger climate classification" (http://www.hydrol-earth-syst-sci.net/11/1633/2007/hess-11-1633-2007.html). Hydrol-earth-syst-sci.net. 2010-07-04. . Retrieved 2010-08-18.

[23] "GLOBAL ECOLOGICAL ZONING FOR THE GLOBAL FOREST RESOURCES ASSESSMENT 2000" (http://www.fao.org/docrep/006/ad652e/ad652e07.htm). Fao.org. . Retrieved 2010-08-18.

[24] "2011 Sub-Provincial Population Estimates" (http://www.bcstats.gov.bc.ca/data/pop/pop/mun/SubProvincialEstimates_Current.pdf). Government of British Columbia. 2011. . Retrieved 20 January 2012.

[25] "U.S. Census Bureau Delivers Washington's 2010 Census Population Totals" (http://2010.census.gov/news/releases/operations/cb11-cn45.html). United States Census Bureau. February 23, 2011. . Retrieved 11 July 2011.

[26] "Population and Housing Occupancy Status: 2010 - United States -- Metropolitan Statistical Area; and for Puerto Rico" (http://factfinder2.census.gov/faces/tableservices/jsf/pages/productview.xhtml?pid=DEC_10_NSRD_GCTPL2.US24PR&prodType=table). *2010 United States Census*. United States Census Bureau, Population Division. April 14, 2011. . Retrieved 12 July 2011.

[27] "2011 Sub-Provincial Population Estimates" (http://www.bcstats.gov.bc.ca/data/pop/pop/mun/SubProvincialEstimates_Current.pdf). Government of British Columbia. 2011. . Retrieved 20 January 2012.

[28] Part of Vancouver metropolitan area.

[29] "Washington population by county - Census 2010: Washington" (http://data.spokesman.com/census/2010/washington/counties). The Spokesman-Review. . Retrieved 12 July 2011.

[30] Part of Seattle metropolitan area (Seattle-Tacoma-Bellevue, WA MSA).

[31] Sum of the populations of the cities of Kennewick, Pasco, and Richland.

[32] Part of Portland metropolitan area (Portland-Vancouver-Hillsboro, OR-WA MSA).

[33] "Census 2006 Community Profiles: Abbotsford, City and CMA" (http://www12.statcan.ca/census-recensement/2006/dp-pd/prof/92-591/details/page.cfm?Lang=E&Geo1=CSD&Code1=5909052&Geo2=CMA&Code2=932&Data=Count&SearchText=Abbotsford&SearchType=Begins&SearchPR=01&B1=All&Custom=). Government of Canada. 2006. . Retrieved 12 July 2011.

[34] "2011 Sub-Provincial Population Estimates" (http://www.bcstats.gov.bc.ca/data/pop/pop/mun/SubProvincialEstimates_Current.pdf). Government of British Columbia. 2011. . Retrieved 20 January 2012.

[35] "2011 Sub-Provincial Population Estimates" (http://www.bcstats.gov.bc.ca/data/pop/pop/mun/SubProvincialEstimates_Current.pdf). Government of British Columbia. 2011. . Retrieved 20 January 2012.

[36] Font size Print E-mail Share (2000-10-25). "Northwest Passage" (http://www.cbsnews.com/stories/2000/10/24/politics/main243670.shtml). CBS News. . Retrieved 2010-08-18.

[37] (http://www.seattle.gov/mayor/about/precedingmayors.htm)

[38] "Communism in Washington State - History and Memory Project" (http://depts.washington.edu/labhist/cpproject/). Depts.washington.edu. . Retrieved 2011-02-19.

[39] "HistoryLink Essay: Group Health Cooperative - Part 1: Planting the Seeds, 1911-1945" (http://www.historylink.org/essays/output.cfm?file_id=7531). Historylink.org. 2005-11-13. . Retrieved 2011-02-19.

[40] Aluminum, Columbia River History (http://www.nwcouncil.org/history/Aluminum.asp), Northwest Power & Conservation Council

[41] Arthur Frommer (2008-12-08). "Vancouver's melting pot makes a terrific cultural stew" (http://www.chicagotribune.com/travel/chi-frommer-vancouver-1207__r_t_dec07,0,6420476.story). *Chicago Tribune*. .

[42] The Republic of Cascadia (http://republic-of-cascadia.tripod.com/) Retrieved on 2009-12-23

[43] "Retrieved on 2009-12-23 [http://www.worldchanging.com/archives/004331.html Retrieved on 2009-12-23" (http://cascadia-bioregion.tripod.com/). Cascadia-bioregion.tripod.com. 1924-05-03. . Retrieved 2011-02-19.

[44] Abbott, Carl (2002). "8. That Long Western Border" (http://books.google.com/books?id=4leTYpWOE5UC&pg=PA203). In Findlay, John M.; Coates, Ken S.. *Parallel Destinies: Canadian-American Relations West of the Rockies*. University of Washington Press. pp. 203–213–5. ISBN 9780295982526. . Retrieved 14 December 2010.

[45] Svoboda, Elizabeth (2010-08-07). "America's 50 Greenest Cities | Popular Science" (http://www.popsci.com/environment/article/2008-02/americas-50-greenest-cities?page=1). Popsci.com. . Retrieved 2010-08-18.

[46] Bower, Craig S. (2008-06-19). "North America's most bike-friendly cities - Travel - Active Travel - msnbc.com" (http://www.msnbc.msn.com/id/25267048/). MSNBC. . Retrieved 2010-08-18.
[47] Gross, Matt (May 10, 2009). "Frugal Portland" (http://travel.nytimes.com/2009/05/10/travel/10Portland.html). *The New York Times*. . Retrieved May 20, 2010.
[48] Frequently Asked Questions - Medical Use of Marihuana (http://www.hc-sc.gc.ca/dhp-mps/marihuana/about-apropos/faq-eng.php), Health Canada
[49] Washington Medical Marijuana Legality (http://apps.leg.wa.gov/RCW/default.aspx?cite=69.51A&full=true), Washington State RCW
[50] "Oregon Medical Marijuana Legality" (http://www.oregon.gov/DHS/ph/ommp/data.shtml). Oregon.gov. 2011-02-01. . Retrieved 2011-02-19.
[51] "Spirit of 12" (http://www.seahawks.com/fans/12zone/spirit-of-12.html). Seahawks.com. . Retrieved 2010-08-18.
[52] http://www.gohuskies.com/facilities/husky-stadium.html
[53] http://www.michigandaily.com/content/j-brady-mccollough-duck-duck-lose
[54] "Seattle Top Gaming City?" (http://news.digitaltrends.com/article10268.html). Digital Trends. May 2, 2006. .
[55] "uwnews.washington.edu" (http://uwnews.washington.edu/ni/article.asp?articleID=2817). uwnews.washington.edu. 1997-01-15. . Retrieved 2011-02-19.
[56] "Can We Still Speak Chinook?" (http://thetyee.ca/Life/2006/01/10/StillSpeakChinook/). Thetyee.ca. . Retrieved 2011-02-19.
[57] "Religion and Public Life in the Pacific Northwest" (http://www.religionatlas.org/religion_region/COASTALNORTHWEST.htm). Religionatlas.org. . Retrieved 2011-02-19.
[58] "Charting the unchurched in America" (http://www.usatoday.com/life/2002/2002-03-07-no-religion.htm). *USA Today*. March 7, 2002. . Retrieved May 20, 2010.
[59] "Religious identification in the U.S" (http://www.religioustolerance.org/chr_prac2.htm). Religioustolerance.org. . Retrieved 2011-02-19.
[60] "American Religious Identification Survey" (http://livinginliminality.files.wordpress.com/2009/03/aris_report_2008.pdf). March 2009. .
[61] Vance, Chris (July 24, 2007). "Crosscut Seattle - Why Washington Republicans got creamed in 2006 and what they can do about it" (http://www.crosscut.com/politics-government/5511/). Crosscut.com. . Retrieved 2011-08-13.
[62] Yardley, William (April 11, 2008). "Dalai Lama Arrives for a Five-Day Conference in Seattle, Very Much His Kind of Town" (http://www.nytimes.com/2008/04/11/us/11seattle.html?_r=1&oref=slogin). *The New York Times*. . Retrieved May 20, 2010.
[63] Mennonite Central Committee Supportive Care Services (http://www.mccscs.com/cms/index.php?module=ContentExpress&func=display&ceid=9)
[64] "ABC News: School Says Halloween Disrespectful to Witches" (http://abcnews.go.com/US/Story?id=184701&page=1). Abcnews.go.com. 2004-10-21. . Retrieved 2011-02-19.
[65] Oregon Historical Society article about Old Believers (http://www.ohs.org/education/oregonhistory/historical_records/dspDocument.cfm?doc_ID=764E6BED-FFC4-C034-9A5563F41CE37080). Retrieved February 9, 2007.

References

Further reading

- Gastil, Raymond D., and Barnett Singer. *The Pacific Northwest: Growth of a Regional Identity* (McFarland, 2010) 221 pp. isbn 978-0-7864-4540-0
- Inglis, Robin. *Historical Dictionary of the Discovery and Exploration of the Northwest Coast of America* (Scarecrow, 2008) lxxvi+429 pp. isbn 978-0-8108-5551-9
- Schwantes, Carlos A. *The Pacific Northwest: An Interpretive History* (2nd ed. 2000)

Confluence

In geography, a **confluence** is the meeting of two or more bodies of water. Known also as a *conflux*,[1] it refers either to the point where a tributary joins a larger river, called the main stem, or where two streams meet to become the source of a river of a new name, such as the confluence of the Monongahela and Allegheny Rivers in Pittsburgh, Pennsylvania creating the Ohio River.

Confluence of rivers, the Mosel flows into the Rhine at Koblenz. The Name "Koblenz" itself has its origin in the latin name "Confluentes"

The term is also used to describe the meeting of tidal or other non-riverine bodies of water, such as two canals[2] or a canal and a lake.[3] A one-mile (1.6 km) portion of the Industrial Canal in New Orleans accommodates the Gulf Intracoastal Waterway and the Mississippi River-Gulf Outlet Canal; therefore those three waterways are confluent there.

Notable confluences

Confluence of the Bhagirathi and Alaknanda Rivers to produce the Ganges at Devprayag, India. Note the swirl of sediment from the Alaknanda.

- The *Sangam, near* Allahabad, India, where the sacred rivers Ganges, Yamuna and Saraswati meet to create one of the holiest places in Hinduism.
- The Ohio River is formed by the confluence of the Monongahela and Allegheny Rivers, located in Pittsburgh, Pennsylvania.
- The Madison, Jefferson and Gallatin Rivers in Three Forks, Montana form the confluence of the Missouri River, one of the longest rivers in the United States (2,341 miles (3,767 km)).
- Kuala Lumpur, the capital of Malaysia, is situated at the confluence of the Gombak River (previously known as Sungai Lumpur, which means muddy river) and the Klang River.
- Passau known also as the Dreiflüssestadt (City of Three Rivers), because the Danube River is joined there by the Inn River from the South, and the Ilz River coming out of the Bavarian Forest to the North.
- Manaus, Brazil is located on the Rio Negro near its confluence with the Amazon. It is the chief port and a hub for the region's extensive river system.
- Osijek, Croatia is located on the right bank of the river Drava 25 km upstream of its confluence with the Danube.
- Belgrade, the capital of Serbia lies at the confluence of the Sava and Danube Rivers.
- The Hochelaga Archipelago, including the island and city of Montreal, is located at the confluence of the St. Lawrence River and Ottawa River in Quebec, Canada.
- The two largest rivers in Australia, the Darling and the Murray converge at Wentworth, New South Wales

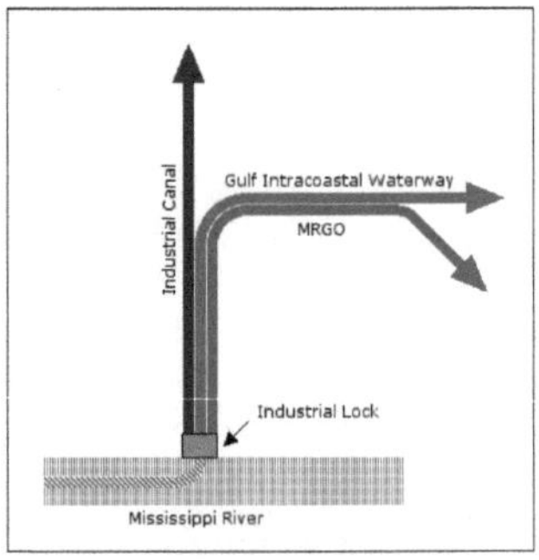

Confluence of canals
This simplified diagram shows how a section of the Industrial Canal in New Orleans also serves as the channel for the Gulf Intracoastal Waterway and the Mississippi River-Gulf Outlet Canal. At the bottom, a portion of the Intracoastal is also shown to be "confluent" with the Mississippi River.

- Rovaniemi, the capital of Finnish Lapland and one of the largest towns above the Arctic Circle, is located on the confluence of rivers Ounasjoki and Kemijoki.
- Winnipeg, Canada is located at the confluence of the Red River, and the Assiniboine River. The area is referred to as The Forks by locals, and has been an important trade location for over 6000 years.
- Triangle of Three Emperors, former tripoint in Europe.
- Cairo, Illinois in the United States where the Ohio River flows into the Mississippi River.
- St. Louis, Missouri is built just south of the confluence of the Mississippi River and the Missouri River.
- Lyon, France is located on the confluence of the Saône and Rhone rivers.
- The confluence between the Franklin River and Gordon River in south-western Tasmania
- The confluence of the Snake and Columbia River at the Tri-Cities of Washington.
- The confluence of the rivers Pivka and Rak is the largest subterranean confluence in Europe.
- The confluence at Lokoja, Nigeria, where the Niger and Benue rivers merge.
- The Sudanese capital of Khartoum, located at the confluence of the White Nile and the Blue Nile. This merging point is the beginning of the Nile.
- Columbus, Ohio in the United States is located at the confluence of the Scioto and Olentangy rivers.

The confluence of the Rio Negro (black) and the Rio Solimões (sandy) near Manaus, Brazil.

See also

- Mouth (water stream)

- Aber and Inver as place-name elements
- Sangam at Allahabad

References

[1] http://www.merriam-webster.com/dictionary/conflux
[2] The U.S. Army Corps of Engineers refers to the confluence of the Assawoman Canal with the Bethany Loop Canal in Delaware. See: "CENAP-OP-R-Quarterly Report, 2004-05-12" (http://www.nap.usace.army.mil/cenap-pa/qtrlyPN-2-2004.htm). *Philadelphia Engineer District.* . Retrieved 2006-03-11.
[3] Engineers in New Orleans refer to the confluence of the 17th Street Canal and Lake Pontchartrain. See: "Interim Closure Structure at 17th St. Canal" (http://web.archive.org/web/20060625180619/http://www.mvn.usace.army.mil/tfg/OEB09.htm). *Task Force Guardian.* Archived from the original (http://www.mvn.usace.army.mil/tfg/OEB09.htm) on 2006-06-25. . Retrieved 2006-03-11.

Willamette River

Willamette River	
 The Willamette passing through downtown Portland	
Name origin: From a Clackamas Indian village name[1] [2]	
Country	United States
State	Oregon
Source	Confluence of Middle Fork Willamette River and Coast Fork Willamette River
- location	near Eugene, Lane County, Oregon
- elevation	438 ft (134 m) [3]
- coordinates	44°01′23″N 123°01′25″W [3]
Mouth	Columbia River
- location	Portland, Multnomah County, Oregon
- elevation	10 ft (3 m) [3]
- coordinates	45°39′10″N 122°45′53″W [3]
Length	187 mi (301 km) [4]
Basin	11478 sq mi (29730 km^2) [5]
Discharge	for Morrison Bridge, Portland, 12.8 miles (20.6 km) from mouth
- average	32950 cu ft/s (900 m^3/s) [6]
- max	420000 cu ft/s (11900 m^3/s)
- min	4200 cu ft/s (100 m^3/s)

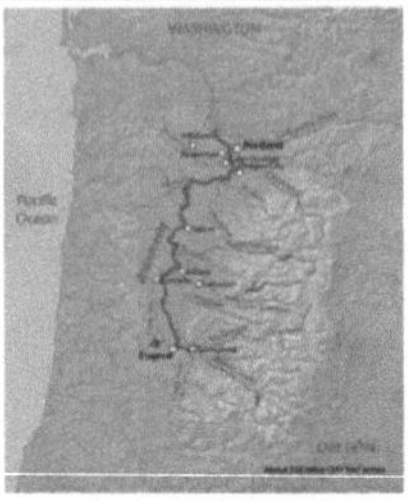

A map of the Willamette River, its drainage basin, major tributaries and major cities

The **Willamette River** (pronounced English pronunciation: /wɨˈlæmɨt/ (listen)) is a major tributary of the Columbia River, accounting for 12 to 15 percent of the Columbia's flow. The Willamette's main stem is 187 miles (301 km) long, lying entirely in northwestern Oregon in the United States. Flowing northward between the Oregon Coast Range and the Cascade Range, the river and its tributaries form the Willamette Valley, a basin that contains two-thirds of Oregon's population, including the state capital, Salem, and the state's largest city, Portland. Portland surrounds the Willamette's mouth at the Columbia.

Originally created by plate tectonics about 35 million years ago and subsequently altered by volcanism and erosion, the river's drainage basin was significantly modified by the Missoula Floods at the end of the most recent ice age. Humans began living in the watershed over 10,000 years ago. There were once many tribal villages along the lower river and in the area around its mouth on the Columbia. Indigenous peoples lived throughout the upper reaches of the basin as well.

Rich with sediments deposited by flooding and fed by prolific rainfall on the western side of the Cascades, the Willamette Valley is one of the most fertile agricultural regions in North America, and was thus the destination of many 19th-century pioneers traveling west along the Oregon Trail. The river was an important transportation route during this time, although Willamette Falls, just upstream from Portland, was a major barrier to boat traffic. In the 21st century, major highways follow the river and roads cross it on more than 50 bridges.

Since 1900, more than 15 large dams and many smaller ones have been built in the Willamette's drainage basin, and 13 of them are managed by the U.S. Army Corps of Engineers (USACE). The dams are used primarily to produce hydroelectricity, to store water for irrigation, and to divert water into deeper, narrower channels in order to prevent flooding. The river and its tributaries support 60 fish species, including many species of salmon and trout; this is despite the dams, other alterations, and pollution (especially on the river's lower reaches). Part of the Willamette Floodplain was established as a National Natural Landmark in 1987 and the river was named as one of 14 American Heritage Rivers in 1998.

Course

The upper tributaries of the Willamette originate in the mountains south and southeast of Eugene, Oregon, and Springfield. Formed by the confluence of the Middle Fork Willamette River and the Coast Fork Willamette River near Springfield, the main stem Willamette meanders generally north for 187 miles (301 km) to the Columbia River. The river's two most significant course deviations occur at Newberg, where it turns sharply east, and about 18 miles (29 km) downstream from Newberg, where it turns north again. Near its mouth north of downtown Portland, the river splits into two channels that flow around Sauvie Island. Used for navigation purposes, these channels are managed by the U.S. federal government. The main channel, which is 40 feet (12 m) deep and varies in width from 600 to 1900 feet (180 to 580 m) (although the river broadens to 2000 feet (610 m) in some of its lower reaches),[7] [8] enters the Columbia about 101 miles (163 km) from the larger river's mouth on the Pacific Ocean. The channel forms the primary navigational conduit for Portland's harbor and riverside industrial areas. The smaller Multnomah Channel, a distributary, is 21 miles (34 km) long, about 600 feet (180 m) wide, and 40 feet (12 m) deep. It ends about 14.5 miles (23.3 km) further downstream on the Columbia, near St. Helens in Columbia County.[7] [8] [9] [10] [11]

Proposals have been made for deepening the Multnomah Channel to 43 feet (13 m) in conjunction with roughly 103.5 miles (166.6 km) of tandem-maintained navigation on the Columbia River.[11] Between the 1850s and the 1960s, channel-straightening and flood control projects, as well as agricultural and urban encroachment, cut the length of the river between the McKenzie River confluence and Harrisburg by 65 percent. Similarly, the river was shortened by 40 percent in the stretch between Harrisburg and Albany.[12]

Interstate 5 and U.S. Route 99 are the two major highways that follow the river for its entire length. Communities along the main stem include Springfield and Eugene in Lane County; Harrisburg in Linn County; Corvallis in Benton County; Albany in Linn and Benton counties; Salem in Marion County; Newberg in Yamhill County; Oregon City, West Linn, Milwaukie, and Lake Oswego in Clackamas County; and Portland in Multnomah and Washington counties. Significant tributaries from source to mouth include the Middle and Coast forks and the McKenzie, Long Tom, Marys, Calapooia, Santiam, Luckiamute, Yamhill, Molalla, Tualatin, and Clackamas rivers.[7] [9]

Beginning at 438 feet (134 m) above sea level, the main stem descends 428 feet (130 m) between source and mouth, or about 2.3 feet per mile (0.4 m per km).[3] The gradient is slightly steeper from the source to Albany than it is from Albany to Oregon City.[13] At Willamette Falls, between West Linn and Oregon City, the river plunges about 40 feet (12 m).[13] For the rest of its course, the river is extremely low-gradient and is affected by Pacific Ocean tidal effects from the Columbia.[13] The main stem of the Willamette varies in width from about 330 to 660 feet (100 to 200 m).[13]

Discharge

With an average flow at the mouth of about **unknown operator: u','** cubic feet per second (**unknown operator: u'strong'unknown operator: u','**m^3/s), the Willamette ranks 19th in volume among rivers in the United States[14] and contributes 12 to 15 percent of the total flow of the Columbia River.[15] The Willamette's flow varies considerably season to season, averaging about 8200 cubic feet per second (230 m^3/s) in August to more than 79000 cubic feet per second (2200 m^3/s) in December.[13]

The Willamette River at Harrisburg

The U.S. Geological Survey (USGS) operates four stream gauges along the river, at Harrisburg, Albany, Salem, and Portland. The average discharge at the lowermost gauge, near the Morrison Bridge in Portland, was 32950 cubic feet per second (933 m^3/s) between 1972 and 2009. Located at river mile (RM) 12.8 or river kilometer (RK) 20.6, the gauge measures the flow from an area of 11200 square miles (29000 km^2), roughly 97 percent of the Willamette basin.[] The highest flow recorded at this station was 420000 cubic feet per second (11893 m^3/s) on February 9, 1996, during the Willamette Valley Flood of 1996, and the minimum was 4200 cubic feet per second (120 m^3/s) on July 10, 1978.[] The highest recorded flow of 635000 cubic feet per second (18000 m^3/s) for the Willamette at a different gauge in Portland occurred during a flood in 1861. This and many other large flows preceded the Flood Control Act of 1936 and dam construction on the Willamette's major tributaries.[16]

The river below Willamette Falls, 26.5 miles (42.6 km) from the mouth, is affected by semidiurnal tides,[17] and gauges have detected reverse flows (backwards river flows) upstream from Ross Island at RM 15 (RK 24).[18] The National Weather Service issues tide forecasts for the river at the Morrison Bridge.[19]

Geology

The Willamette River basin was created primarily by plate tectonics and volcanism and was altered by erosion and sedimentation, including some related to enormous glacial floods as recent as 13,000 years ago.[20] The initial trough-like configuration was created about 35 million years ago[15] as a forearc basin while the Pacific Plate subducted beneath the North American Plate.[20] Marine deposits on top of older volcanics underlie the valley, which was initially part of the continental shelf, rather than a separate inland sea.[21] About 20 to 16 million years ago, uplift formed the Coast Range and separated the basin from the Pacific Ocean.[20]

Basalts of the Columbia River Basalt Group, from eruptions in eastern Oregon, flowed across large parts of the northern half of the basin about 15 million years ago.[20] They covered the Tualatin Mountains (West Hills), most of the Tualatin Valley, and the slopes of hills further south, with up to 1000 feet (300 m) of lava.[22] Later depositions covered the basalt with up to 1000 feet (300 m) of silt in the Portland and Tualatin basins.[23] During the Pleistocene, beginning roughly 2.5 million years ago, volcanic activity in the Cascades combined with a cool, moist climate to produce further heavy sedimentation across the basin; braided rivers created alluvial fans spreading down from the east.[20]

The glacial Bellevue Erratic at Erratic Rock State Natural Site. The rock was transported to the Willamette Valley by the Missoula Floods.[24]

Between about 15,500 and 13,000 years ago, the Missoula Floods—a series of large outpourings originating at Glacial Lake Missoula in Montana—swept down the Columbia River and backfilled the Willamette watershed.[20] Each flood produced "discharges that exceeded the annual discharge of all the present-day rivers of the world combined".[15] Filling the Willamette basin to depths of 400 feet (120 m) in the Portland region, each flood created a temporary lake, Lake Allison, that stretched from Lake Oswego to near Eugene.[25] The ancestral Tualatin Valley, part of the Willamette basin, flooded as well; water depths ranged from 200 feet (61 m) at Lake Oswego to 100 feet (30 m) as far upstream (west) as Forest Grove.[25] Flood deposits of silt and clay, ranging in thickness from 115 feet (35 m) in the north to about 15 feet (4.6 m) in the south, settled from this muddy water to form today's valley floor.[20] The floods carried Montana icebergs well into the basin, where they melted and dropped glacial erratics on the land's surface. These rocks, composed of granite and other materials common to central Montana but not to the Willamette Valley, include more than 40 boulders, each at least 3 feet (0.9 m) in diameter.[26] Before being partly chipped away and removed, the largest of these originally weighed about 160 short tons (150 t).[26]

The northern part of the watershed is underlain by a network of faults capable of producing earthquakes at any time, and many small quakes have been recorded in the basin since the mid-19th century.[27] In 1993, the Scotts Mills earthquake—the largest recent earthquake in the valley, measuring 5.6 on the Richter scale—was centered near Scotts Mills, about 34 miles (55 km) south of Portland.[28] It caused $30 million in damage, including harm to the Oregon State Capitol in Salem.[29] Evidence suggests that massive quakes of 8 or more on the Richter scale have occurred historically in the Cascadia subduction zone off the Oregon coast, most recently in 1700 CE, and that others as strong as 9 on the Richter scale occur every 500 to 800 years.[30] The basin's high population density, its nearness to this subduction zone, and its loose soils, which tend to amplify shaking, make the Willamette Valley especially vulnerable to damage from strong earthquakes.[27]

Watershed

The Willamette River drains a region of 11478 square miles (29730 km^2), which is 12 percent of the total area of Oregon.[5] Bounded by the Coast Range to the west and the Cascade Range to the east, the river basin is about 180 miles (290 km) long and 100 miles (160 km) wide.[15] Elevations within the watershed range from 10495 feet (3199 m) at Mount Jefferson in the Cascade Range[15] to 10 feet (3.0 m) at the mouth on the Columbia River.[3] Watersheds bordering the Willamette River basin are those of the Little Deschutes River to the southeast, the Deschutes River to the east, and the Sandy River to the northeast; the North Umpqua and Umpqua rivers to the south; coastal rivers including (from south to north) the Siuslaw, the Alsea, the Yaquina, the Siletz, the Nestucca, the Trask, and the Wilson to the west; the Nehalem and the Clatskanie to the northwest, and the Columbia River to the north.[31]

About 2.5 million people lived in the Willamette River basin as of 2010, about 65 percent of the population of Oregon.[32] As of 2009, the basin contained 20 of the 25 most populous cities in Oregon.[7] [33] These cities include Springfield, Eugene, Corvallis, Albany, Salem, Keizer, Newberg, Oregon City, West Linn, Milwaukie, Lake Oswego, and Portland.[7] [33] The largest is Portland, with more than 500,000 residents.[33] Other cities in the watershed (but not on the main-stem river) with populations of 20,000 or more are Gresham, Hillsboro, Beaverton, Tigard, McMinnville, Tualatin, Woodburn, and Forest Grove.[7] [33]

Sixty-four percent of the watershed is privately owned, while 36 percent is publicly owned.[34] The U.S. Forest Service manages 30 percent of the watershed, the U.S. Bureau of Land Management 5 percent, and the State of Oregon 1 percent.[34] Sixty-eight percent of the watershed is forested; agriculture, concentrated in the Willamette Valley, makes up 19 percent, and urban areas cover 5 percent.[34] More than 81000 miles (km) of roads criss-cross the watershed.[34]

In 1987, the U.S. Secretary of the Interior designated 713 acres (289 ha) of the watershed in Benton County as a National Natural Landmark. This area is the Willamette Floodplain, the largest remaining unplowed native grassland in the North Pacific geologic province, which encompasses most of the Pacific Northwest coast.[35]

History

First inhabitants

The Willamette River near the confluence with the Molalla River

For at least 10,000 years, a variety of indigenous peoples populated the Willamette Valley. These included the Kalapuya, the Chinook, and the Clackamas.[36] The territory of the Clackamas encompassed the northeastern portion of the basin, including the Clackamas River (with which their name is shared). Although it is unclear exactly when, the territory of the Chinook once extended across the northern part of the watershed, through the Columbia River valley. Indigenous peoples of the Willamette Valley were further divided into groups including the Yamhill, Atfalati (Tualatin), Molala, Santiam, Chuchsney-Tufti, Muddy Creek, Long Tom, Yoncalla, Siuslaw, Calapooia, Chepenafa and Luckiamute, many of which were part of the larger Kalapuya group.[5] [36] The name *Willamette* is of indigenous origin, deriving from the French pronunciation of the name of a Clackamas Native American village.[1] However, Native American languages in Oregon were very similar,[37] so the name may also be derived from Kalapuya dialects.[38]

Around the year 1850, the Kalapuya numbered between 2,000 and 3,000 and were distributed among several groups. These figures are only speculative; there may have been as few as eight subgroups or as many as 16.[39] In that time period, the Clackamas' tribal population was roughly 1,800.[40] The U.S. Census Bureau estimated that the Chinook population was nearly 5,000,[41] though not all of the Chinook lived on the Willamette. The Chinook territory encompassed the lower Columbia River valley and significant stretches of the Pacific coast on both the north and the south side of the Columbia's mouth. At times, however, the Chinook territory extended even farther south in the Willamette Valley.[42] The total native population was estimated at about 15,000.[36]

The indigenous peoples of the Willamette River practiced a variety of life ways. Those on the lower river, slightly closer to the coast, often relied on fishing as their primary economic mainstay. Salmon was the most important fish to Willamette River tribes as well as to the Native Americans of the Columbia River, where white traders traded fish with the Native Americans. Upper-river tribes caught steelhead and salmon, often by building weirs across tributary streams. Tribes of the northern Willamette Valley practiced a generally settled lifestyle. The Chinooks lived in great wooden lodges,[43] practiced slavery, and had a well-defined caste system.[43] People of the south were more nomadic, traveling from place to place with the seasons. They were known for the controlled burning of woodlands to create meadows for hunting and plant gathering (especially camas).[44]

Fur trade

The Willamette River first entered the records of outsiders in 1792, when it was seen by British Lieutenant William Robert Broughton of the Vancouver Expedition, led by George Vancouver.[45] From the 18th to the mid-19th centuries, much of the Pacific Northwest and most of its rivers were involved in the fur trade, in which fur trappers (mostly French-Canadians working for the Hudson's Bay Company and the North West Company, which later merged) hunted for beaver and sea otter on rivers, streams, and coastlines. The pelts of these animals commanded substantial prices in either the United States, Canada or eastern Asia, because of their "thick, luxurious and water-repellent" qualities.[46]

Fur traders heavily exploited the Willamette River and its tributaries.[47] During this period, the Siskiyou Trail (or California-Oregon Trail) was created. This trading path, over 600 miles (970 km) long, stretched from the mouth of the Willamette River near present-day Portland south through the Willamette Valley, crossing the Cascades and the Siskiyou Mountains, and south through the Sacramento Valley to San Francisco, California.[48]

19th-century development

In 1805, the Lewis and Clark Expedition traveled thousands of miles across central North America in an attempt to map and explore the Louisiana Territory of the United States and the Oregon Country, which were then occupied mainly by Native Americans and settlers from Great Britain. As the expedition traveled down and back up the Columbia River, it missed the mouth of the Willamette, one of the Columbia's largest tributaries. It was only after receiving directions from natives along the Sandy River that the explorers learned about their oversight. William Clark returned down the Columbia and entered the Willamette River in April 1806.[15] The United States Exploring Expedition passed through the Willamette Valley in 1841 while traveling along the Siskiyou Trail. The expedition members noted extensive salmon fishing by natives at Willamette Falls, much like that at Celilo Falls on the Columbia River.[49]

In the middle part of the 19th century, the Willamette Valley's fertile soils, pleasant climate, and abundant water attracted thousands of settlers from the eastern United States, mainly the Upland South borderlands of Missouri, Iowa, and the Ohio Valley.[50] Many of these emigrants followed the Oregon Trail, a 2170-mile (3490 km) trail across western North America that began at Independence, Missouri, and ended at various locations near the mouth of the Willamette River. Although people had been traveling to Oregon since 1836, large-scale migration did not begin until 1843, when nearly 1,000 pioneers headed westward. Over the next 25 years, some 500,000 settlers traveled the Oregon Trail, braving the rapids of the Snake and Columbia Rivers to reach the Willamette Valley.[51] [52] [53]

Oregon City circa 1867, with Willamette Falls in the background

Starting in the 1820s, Oregon City developed near Willamette Falls. It was incorporated in 1844, becoming the first city west of the Rocky Mountains to have that distinction.[54] John McLoughlin, a Hudson's Bay Company (HBC) official, was one of the major contributors to the founding of the town in 1829.[55] McLoughlin attempted to persuade the British government (which still held sway over the area) to allow American settlers to live on the land, and provided significant help to American colonization of the area, all against the HBC's orders.[56] Oregon City prospered because of the paper mills that were run by the water power of Willamette Falls, but the falls formed an impassable barrier to river navigation. Linn City (originally Robins Nest) was established across the Willamette from Oregon City.[57]

After Portland was incorporated in 1851, quickly growing into Oregon's largest city, Oregon City gradually lost its importance as the economic and political center of the Willamette Valley. Beginning in the 1850s, steamboats began to ply the Willamette, despite the fact that they could not pass Willamette Falls.[58] As a result, navigation on the Willamette River was divided into two stretches: the 27-mile (43 km) lower stretch from Portland to Oregon City—which allowed connection with the rest of the Columbia River system—and the upper reach, which encompassed most of the Willamette's length.[59] Any boats whose owners found it absolutely necessary to get past the falls had to be portaged. This led to competition for business among steam portage companies.[60] [61] In 1873, the construction of the Willamette Falls Locks bypassed the falls and allowed easy navigation between the upper and lower river. Each lock chamber measured 210 feet (64 m) long and 40 feet (12 m) wide, and the canal was originally operated manually before it switched to electrical power.[61] Today, the lock system is little used.[62]

As commerce and industry flourished on the lower river, most of the original settlers acquired farms in the upper Willamette Valley. By the late 1850s, farmers had begun to grow crops on most of the available fertile land.[63] The settlers increasingly encroached on Native American lands. Skirmishes between natives and settlers in the Umpqua and Rogue valleys to the southwest of the Willamette River led the Oregon state government to remove the natives by military force.[64] They were first led off their traditional lands to the Willamette Valley, but soon were marched to the Coast Indian Reservation. In 1855, Joel Palmer, an Oregon legislator, negotiated a treaty with the Willamette Valley tribes, who, although unhappy with the treaty, ceded their lands to non-natives.[65] [66] The natives were then relocated by the government to a part of the Coast Reservation that later became the Grande Ronde Reservation.[66]

Between 1879 and 1885, the Willamette River was charted by Cleveland S. Rockwell, a topographical engineer and cartographer for the U.S. Coast and Geodetic Survey. Rockwell surveyed the lower Willamette from the foot of Ross Island through Portland to the Columbia River and then downstream on the Columbia to Bachelor Island.[67] Rockwell's survey was extremely detailed, including 17,782 hydrographic soundings. His work helped open the port of Portland to commerce.[68]

In the second half of the 19th century, the USACE dug deep channels and built locks and levees in the Willamette's watershed. These navigational advantages permitted businesses to deliver lumber and agricultural products to Portland, feeding the city's growing economy. Trade goods from afar could also reach settlements in the lower Willamette Valley, via the channels at the Willamette's mouth on the Columbia. However, lumber was also transported on a network of railroads that ran from mills to major settlements.[69]

20th and 21st centuries

Conveyor belt loading debris onto a barge as part of the Big Pipe Project

By the early 20th century, major river-control projects had begun to take place. Levees were constructed along the river in most urban areas, and Portland built concrete "seawalls" to protect its downtown sector.[16] In the following decades, many large dams were built on Cascade Range tributaries of the Willamette. The Army Corps of Engineers operates 13 such dams, which affect flows from about 40 percent of the basin.[70] Most of them do not have fish ladders.[71]

With development in and near the river came increased pollution. By the late 1930s, efforts to stem the pollution led to formation of a state sanitary board to oversee modest cleanup efforts.[5] In the 1960s, Oregon Governor Tom McCall led a push for stronger pollution controls on the Willamette.[72] To enhance historical, natural, and recreational values along the river, the Oregon State Legislature established the Willamette Greenway program in 1967. Through it, state and local governments cooperated in creating or improving a system of parks, trails, and wildlife refuges along the river.[45] In 1998, the Willamette became one of 14 rivers designated an American Heritage River by former U.S. President Bill Clinton.[73] By 2007 the Greenway had grown to include more than 170 separate land parcels, including 10 state parks.[45] Public uses of the river and land along its shores include camping, swimming, fishing, boating, hiking, bicycling, and wildlife viewing.[45]

A 1991 agreement between the City of Portland and the State of Oregon to dramatically reduce combined sewer overflows (CSOs),[74] led to Portland's "Big Pipe Project". The long-term project has been separating the city's sanitary sewer lines from storm-water inputs that sometimes overwhelm the combined system during heavy rains. When that occurs, some of the raw sewage in the system flows into the river instead of into the city's wastewater treatment plant. The City of Portland predicts that Big Pipe and related work will reduce CSO volume on the lower river by more than 94 percent by the end of 2011.[75] [76]

Dams and bridges

Dams

At 463 feet (141 m) high, Detroit Dam is the largest dam in the Willamette River basin.

There are more than 15 major dams on the Willamette's tributaries as well as a complex series of levees, dikes, and channels to control the river's flow. The only dam on the Willamette's main stem, however, is the Willamette Falls Dam, a low weir-type structure at Willamette Falls that diverts water into the headraces of the adjacent mills and a power plant.[77] The dams on the Willamette's major tributaries are primarily large flood-control, water storage, and power-generating dams. Thirteen of these dams were built by the U.S. Army Corps of Engineers (USACE), and 11 of them produce hydropower. Most of the other ones are owned by state or local interests.[78]

On the main stem, numerous flow regulation structures force the river into a narrower and deeper channel to facilitate navigation and flood control.[45] Flood-control dams operated by the USACE are estimated to hold up to 27 percent of the Willamette's runoff. They are used to regulate river flows so as to cut peaks off floods and increase low flows in late summer and autumn.[45] [70] [79]

Large dams and their reservoirs include Hills Creek, Lookout Point, Dexter, and Fall Creek on the Middle Fork and its tributaries; Cottage Grove on the Coast Fork; Cougar on the South Fork McKenzie; Fern Ridge on the Long Tom River; Green Peter, Foster, Detroit, and Big Cliff on various forks of the Santiam; and North Fork, Faraday, Oak

Grove, and Timothy on the Clackamas and its forks.[77] [80] Detroit Dam on the North Santiam River, by far the largest, is 463 feet (141 m) high and stores 455000 acre feet (m^3) of water.[81] USACE dam facilities in the basin are collectively known as the Willamette River Basin Project.[82]

Bridges

The Hawthorne Bridge is the oldest remaining highway structure over the Willamette.

The 50 or so crossings of the Willamette River include many historic structures, such as the Van Buren Street Bridge, a swing bridge. Built in 1913, it carries Oregon Route 34 (Corvallis–Lebanon Highway) over the river upstream of RM 131 (RK 211) in Corvallis. The machinery to operate the swing span was removed in the 1950s.[83] [84] The Oregon City Bridge, built in 1922, replaced a suspension span constructed at the site in 1888. It carries Oregon Route 43 over the river at about RM 26 (RK 42) between Oregon City and West Linn.[85] [86]

The Ross Island Bridge, which carries U.S. Route 26 (Mount Hood Highway) over the river at RM 14 (RK 23), is the furthest upstream of 10 highway bridges that cross the Willamette River in Portland's central business district. The 3700-foot (1100 m) bridge is the only cantilevered deck truss in Oregon.[87] [88] The oldest remaining highway structure over the Willamette is the 1910-built Hawthorne Bridge, located slightly downstream from the Ross Island Bridge.[89]

The lower deck of the Steel Bridge can be raised independently of the upper deck.

Another historic structure, the Steel Bridge, further downstream, was "the largest telescoping bridge in the world at the time of its opening" in 1912.[90] It carries trains on its lower deck, MAX (Metropolitan Area Express) light-rail trains and motorized vehicles on its upper deck, and foot and bicycle traffic on a cantilevered walkway attached to the lower deck.[91] When small ships must pass under the bridge, its double vertical-lift span can raise a lower railway deck without disturbing traffic on the upper deck. Operators can raise both decks as high as 163 feet (50 m) above the water. The Steel Bridge is "believed to be the world's only double-lift span that can raise its lower deck independently of the upper deck."[90]

The Broadway Bridge, slightly downstream of the Steel Bridge, was the world's longest double-leaf bascule drawbridge at the time of its construction in 1913.[92] Further downstream, the St. Johns Bridge, a steel suspension bridge built in 1931, replaced the last of the Willamette River ferries in Portland.[93] At about RM 6 (RK 10), it carries U.S. Route 30 Bypass.[94]

Flooding

Due to the volume and seasonality of precipitation in its valley, the Willamette River has often flooded. In 1861, rainstorms and warm temperatures combined with a well-above-average snowpack in the Cascades created the largest Willamette River flood in recorded history. An observer of the flood wrote, "The whole Willamette valley [*sic*] was a sheet of water".[95] From Eugene to Portland, thousands of acres of rich riverside farmland were washed away and many towns in the valley were damaged or destroyed. The *Great Flood*, as it is sometimes called, was massively destructive to human development because most of that development was located in the river's floodplain. Floodplain locations offer easy access to river transportation and the best soils for farming. The 1861 flood peaked at 635000 cubic feet per second (18000 m^3/s)—which was more than the flow of the Mississippi River—and inundated some 353000 acres (1430 km^2) of land.[16] [95]

In the summer of 1866, the Willamette was measured at 21 feet (6.4 m) above the "low water mark," and there were more flooding worries. This raised the level of the Columbia River and nearly flooded The Dalles, a city on the Columbia.[96]

An aerial view of the 1996 flooding

The floods returned in early 1890, when the river first rose very quickly and then fell very quickly.[97] [98] Portland's main street was completely submerged. Communication over the Cascades was cut off, and many rail lines were forced to shut down.[98] In 1894, another major flood occurred on the Willamette, and although it too caused huge damage, it was not quite as large as that of 1861.[16]

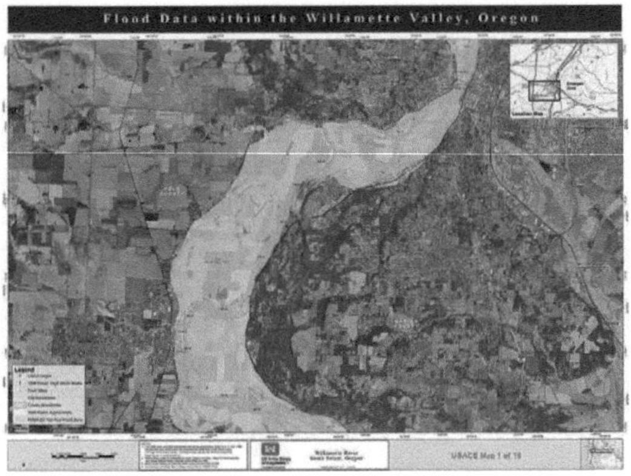

Diagram of the flooding near Salem in 1996

In the 1940s the Willamette once again flooded its valley. It destroyed five bridges in Lane County in December 1942,[99] caused seven deaths in Portland and forced people in Eugene to evacuate in January 1943,[100] [101] caused minor flooding in Corvallis in November 1946,[102] and nearly flooded parts of Salem in December 1948.[103]

Although the Willamette was, by mid-century, heavily engineered and controlled by a complex system of dams, channels, and barriers, it experienced severe floods through the end of the century. Storms caused a major flood that swelled the Willamette and other rivers in the Pacific Northwest from December 1964 through January 1965, submerging nearly 153000 acres (620 km^2) of land.[104] Before dawn on December 21, 1964, the Willamette reached 29.4 feet (9.0 m), which was higher than the Portland seawall. By this time, about 15 people had died as a result of the flooding and about 8,000 Oregonians had been forced to evacuate their homes in search of other shelter.[105]

On December 24, President Lyndon B. Johnson ordered federal aid for the flooded areas. Meanwhile, the Willamette continued to rise.[106] In the next couple of days, the river receded, but on December 27, it was at 29.8 feet (9.1 m), which was still nearly 12 feet (3.7 m) above the flood stage.[107] The Willamette continued to pose flooding threats through January 1965,[108] and more stormy weather occurred along the Pacific Coast.[109]

> "The river crested at one town after another—at Corvallis 3½ feet above flood stage, Oregon City 18 feet above, Portland 10.5 feet above—much like a meal moving through a boa constrictor."
>
> —Associated Press journalist, February 10, 1996, [110]

Very heavy warm rains driven by a subtropical jet stream fell on a high, water-heavy snowpack in the Willamette watershed in February 1996. These conditions, similar to those that caused the 1861 flood, caused some of the costliest flooding in the river's recorded history.[111] An Associated Press journalist wrote, "The river crested at one town after another—at Corvallis 3½ feet above flood stage, Oregon City 18 feet above, Portland 10.5 feet above—much like a meal moving through a boa constrictor."[110] The flood interrupted the progress of Oregon's growing economy.[112] However, the inundated acreage was even smaller than in 1964—only about 117000 acres (470 km^2).[104]

In the wake of the flood in April 1996, about 450 concrete flood-protection walls that had been built in Portland during the flood were removed. Each wall weighed roughly 5500 pounds (2500 kg).[113] In October, however, a new wall was built near Portland, costing about $300,000. It featured 6.35-millimetre (0.250 in) removable steel plates designed to prevent future flooding.[114]

Pollution

Since as early as 1869, with the introduction of a federally-funded "snag puller" designed to keep the waterway clear, human habitation has affected the ecology of the river basin. The construction of large federal dams on the Willamette's tributaries between 1941 and 1969 damaged the spawning grounds for spring Chinook salmon and steelhead trout. Domestic and industrial waste from the cities built up along the river, "essentially turning the main-stem river into an open sewer by the 1920s."[5]

The Tom McCall Waterfront Park is named after an Oregon governor who led a cleanup of the river.

A 1927 Portland City Club report labeled the waterway "filthy and ugly," and identified the City of Portland as the worst offender. The Oregon Anti-Stream Pollution League brought a pollution-abatement measure before the 39th Oregon Legislative Assembly in 1937. The bill passed, but Governor Charles Martin vetoed it. The Izaak Walton League and the Oregon Wildlife Federation countered the governor's veto with a ballot initiative, which passed in November 1938.[5]

Shortly after he was elected in 1966, Governor Tom McCall ordered water quality tests on the Willamette, conducted his own research on the water quality, and became head of the Oregon State Sanitary Authority. McCall learned that the river was heavily polluted in Portland. In a television documentary, "Pollution in Paradise", he said that "the Willamette River was actually cleaner when the Oregon Sanitary Authority was created in 1938 than it was in 1962."[72] He then discouraged tourism in the state and made it harder for companies to qualify for a permit to operate near the river. He also regulated how much those companies could pollute and closed plants that did not meet state pollution standards.[115] [116]

Despite earlier cleanup efforts, state studies in the 1990s identified a wide variety of pollutants in the river bottom, including heavy metals, polychlorinated biphenyls (PCBs), and pesticides along the lower 12 miles (19 km) of the river, in Portland.[117] As a result, this section of the river was designated a Superfund site in 2000,[118] involving the U.S. Environmental Protection Agency (EPA) in cleanup of the river bottom.[119] The initial cleanup is focused on the portion between Swan Island and Sauvie Island.[120] Pollution is exacerbated by combined sewer overflows, which the city is greatly reducing through its Big Pipe Project.[121] Further upstream, the pressing environmental issues have mainly been variations in pH and dissolved oxygen.[122] The Willamette is nevertheless clean enough to be used by communities such as Tigard for drinking water.[123]

Since pollution concerns are primarily along the lower river, the Willamette in general scores relatively high on the Oregon Water Quality Index (OWQI), which is compiled by the Oregon Department of Environmental Quality (DEQ). The DEQ considers index scores of less than 60 to be very poor; the other categories are 60–79 (poor); 80–84 (fair); 85–89 (good), and 90–100 (excellent).[124] The Willamette River's water quality is rated excellent near the source, though it gradually declines to fair near the mouth. Between 1998 and 2007, the average score for the upper Willamette at Springfield (RM 185, RK 298) was 93. At Salem (RM 84, RK 135), the score was 89, and good scores continued all the way to the Hawthorne Bridge in Portland (RM 13, RK 21) at 85. Scores were in the "fair" category further downstream; the least favorable reading was at the Swan Island Channel midpoint (RM 0.5, RK 0.8) at 81. By comparison, sites on the Winchuck River, the Clackamas, and the North Santiam all scored 95, and a site at a pump station on Klamath Strait Drain between Upper Klamath Lake and Lower Klamath Lake recorded the lowest score in Oregon at 19.[125]

Flora and fauna

Over the past 150 years, a significant change for the Willamette River has been the loss of its floodplain forests, which covered an estimated 89 percent of a 400-foot (120 m) band along each river bank in 1850.[126] By 1990 only 37 percent of this zone was forested; the rest had been converted to farm fields or cleared for urban or suburban uses.[126] The remaining forests close to the river include large stands of black cottonwood, Oregon ash, willow, and bigleaf maple.[126] The central valley—a former perennial grass prairie interspersed with oaks, ponderosa pines, and other trees—is devoted almost entirely to farming. Douglas fir, western hemlock, and western red cedar dominate the forest on the Coast Range side of the basin. Forests to the east in the Cascade Range include Pacific silver fir as well as western hemlock and western red cedar.[13]

Osprey are among bird species often seen along the Willamette River.

Fish in the Willamette basin include 31 native species, among them cutthroat, bull, and rainbow trout, several species of salmon, sucker, minnow, sculpin, and lamprey, as well as sturgeon, stickleback, and others. Among the 29 non-native species in the basin, there are brook, brown, and lake trout, largemouth and smallmouth bass, walleye, carp, bluegill, and others. In addition to fish, the basin supports 18 species of amphibians, such as the Pacific giant salamander. Beaver and river otter are among 69 mammal species living in the watershed, also frequented by 154 bird species, such as the American Dipper, Osprey, and Harlequin Duck. Garter snakes are among the 15 species of reptiles found in the basin.[127]

Species diversity is greatest along the lower river and its tributaries. Threatened, endangered, or sensitive species include spring Chinook salmon, winter steelhead, Oregon chub, chum salmon, and Coho salmon.[127] In the central valley, there have been several projects that have restored and protected wetlands[128] in order to provide and preserve habitat for Bald Eagles, Fender's blue butterfly (of which 6,000 remain), Oregon chub, Bradshaw's desert parsley, Willamette Daisy, and Kincaid's lupine.[129] [130] In the early 21st century, osprey populations are increasing along the river, possibly because of a ban on the pesticide DDT and on the birds' ability to use power poles for nesting.[131] Beaver populations, presumed to be much lower than historic levels, are increasing throughout the basin.[131]

The Willamette River as it passes through Portland, Oregon.

See also

- List of crossings of the Willamette River
- List of longest streams of Oregon
- List of rivers of Oregon
- Steamboats of the Willamette River

Notes and references

Notes

[1] Bright, p. 567

[2] See also here.

[3] "Willamette River" (http://geonames.usgs.gov/pls/gnispublic/f?p=gnispq:3:::NO::P3_FID:1158060). Geographic Names Information System, U.S. Geological Survey. November 28, 1980. . Retrieved June 25, 2010. Source and mouth elevation derived from Google Earth search using GNIS source and mouth coordinates.

[4] "Physical Character" (http://www.willamette-riverkeeper.org/WRK/physicalcharacter.html). Willamette Riverkeeper. . Retrieved June 14, 2011.

[5] Robbins, William G. "Willamette River" (http://www.oregonencyclopedia.org/entry/view/willamette_river/). *The Oregon Encyclopedia*. Portland State University. . Retrieved June 4, 2010.

[6] These are the average, maximum, and minimum flows recorded during the period of record (1973–2009) for this gauge. "Water-Data Report 2010: 14211720 Willamette River at Portland, OR" (http://wdr.water.usgs.gov/wy2010/pdfs/14211720.2010.pdf) (PDF). U.S. Geological Survey. pp. 1, 3. . Retrieved July 13, 2011. The greatest recorded Willamette flood, which occurred in December 1861, produced a much larger maximum of 635000 cubic feet per second (18000 m^3/s) in Portland.

[7] DeLorme Mapping. *Oregon Atlas & Gazetteer* (Map) (1991 ed.). Section 41–43, 47–48, 53, 59–60, 66. ISBN 0-89933-235-8.

[8] Bureau of Planning and Sustainability. "Willamette River Dredged Material Management Plan Project" (http://www.portlandonline.com/bps/index.cfm?a=145503&c=42601) (PDF). City of Portland. . Retrieved June 7, 2010.

[9] U.S. Geological Survey. "United States Geological Survey Topographic Map" (http://www.topoquest.com/map.php?lat=44.03988&lon=-123.02511&datum=nad83&zoom=4&map=auto&coord=d&mode=zoomin&size=m). TopoQuest. . Retrieved June 29, 2010. Relevant map quadrangles include Eugene East, Coburg, Junction City, Harrisburg, Peoria, Riverside, Lewisburg, Albany, Monmouth, Salem West, Mission Bottom, Dayton, St. Paul, Newberg, Sherwood, Canby, Lake Oswego, Portland, Linnton, Sauvie Island, and Saint Helens.

[10] "Multnomah Channel" (http://www.oregon.gov/OSMB/library/docs/WillametteRiverGuide-07/WillametteRiverGuidePDF-MultnomahChannel.pdf?ga=t) (PDF). State of Oregon. p. 30. . Retrieved September 14, 2011.

[11] "Background – Project" (http://www.sei.org/columbia/background_project.html). *Columbia River Project—Columbia River Channel Improvement Reconsultation Project*. Sustainable Ecosystems Institute. February 28, 2001. . Retrieved June 7, 2010.

[12] Laenen and Dunnette, p. 24

[13] Benke, *et al.*, p. 617

[14] Kammerer, J.C. (May 1990). "Largest Rivers in the United States" (http://pubs.usgs.gov/of/1987/ofr87-242/). U.S. Geological Survey. . Retrieved July 2, 2010.

[15] Benke, *et al.*, p. 616

[16] "Willamette River Multi-Purpose Development in the 1930s" (http://140.194.76.129/publications/misc/un24/c-8.pdf) (PDF). *Army Engineers and the Development of Oregon: A History of the Portland District U.S. Army Corps of Engineers*. U.S. Army Corps of Engineers. 1983. p. 107. . Retrieved June 6, 2010.

[17] "Tidally-Influenced Waterways" (http://www.oregon.gov/DSL/NAV/tidally.shtml). *Public Ownership of Submerged and Submersible Land*. State of Oregon. . Retrieved July 2, 2010.

[18] Caldwell, James M.; Doyle, Micelis C. "Sediment Oxygen Demand in the Lower Willamette River, Oregon, 1994" (http://or.water.usgs.gov/pubs_dir/Pdf/95-4196.pdf) (PDF). Oregon Department of Environmental Quality. p. 2. . Retrieved July 2, 2010.

[19] National Weather Service Forecast Office, Portland, Oregon (2008). "Local Tide Predictions and Water Levels" (http://www.wrh.noaa.gov/pqr/tides.php). U.S. Department of Commerce. . Retrieved July 2, 2010.

[20] Wallick, Jennifer Rose; Grant, Gordon E.; Lancaster, Stephen T.; Bolte, John P.; Denlinger, Roger P. (2007). "Patterns and Controls on Historical Channel Change in the Willamette River, Oregon, USA" (http://www.fsl.orst.edu/wpg/pubs/08_Wallicketal_Chapter23.PDF) (PDF). Oregon State University. pp. 495–97. . Retrieved August 1, 2010.

[21] Orr, p. 203

[22] Orr, p. 206

[23] Orr, p. 208
[24] Hill, Richard L. (November 29, 2000). "Project takes on story of region's ice-age cataclysms". *The Oregonian* (Portland, Oregon): p. A19.
[25] Orr, pp. 212–13
[26] Orr, p. 214
[27] Orr, pp. 215–19
[28] Ludwin, R.S.; Qamar, A.I.; Malone, S.D.; Crosson, R.S.; Moran, S.; Thomas, G.C., and Steele, W.P. (October 11, 1996). "Scotts Mills Earthquake Summary" (http://spike.geophys.washington.edu/SEIS/EQ_Special/ScottsMills/scottsmills.html). University of Washington Department of Earth and Space Sciences. . Retrieved July 31, 2010.
[29] "Earthquake Education: Historic Earthquakes in the Pacific Northwest" (http://www.oregongeology.com/sub/earthquakes/HistoricEQs.htm). Oregon Department of Geology and Mineral Industries. . Retrieved July 31, 2010.
[30] "Earthquakes" (http://www.ci.troutdale.or.us/documents/hazardmitigationplan/Troutdale_CH-10_Earthquake_10-24-10.pdf) (PDF). City of Troutdale. October 24, 2010. pp. 1, 3. . Retrieved August 31, 2011.
[31] Loy, *et al.*, pp. 168–69
[32] "The basins: about the Willamette River basin" (http://pnwho.forestry.oregonstate.edu/site/index.php). *The Pacific Northwest Hydrologic Observatory*. Oregon State University. . Retrieved August 25, 2011.
[33] "Oregon – Place 2009 Population Estimates" (http://factfinder.census.gov/servlet/GCTTable?_bm=y&-geo_id=04000US41&-_box_head_nbr=GCT-T1-R&-ds_name=PEP_2009_EST&-_lang=en&-format=ST-9S&-_sse=on). U.S. Census Bureau. . Retrieved August 1, 2010.
[34] Benke, *et al.*, pp. 616–17
[35] "Willamette Floodplain" (http://www.nature.nps.gov/nnl/site.cfm?Site=WIFL-OR). National Park Service. March 31, 2011. . Retrieved April 11, 2011.
[36] Spores, Ronald (1993). "Too Small a Place: The Removal of the Willamette Valley Indians, 1850–1856" (http://www.questia.com/googleScholar.qst;jsessionid=MN2LzygdFjyPMTww16TGBbvZQmHblWn3Jw7mHFZP91v0tLKP1WMp!1748356985!-2083947951?docId=96413754). *American Indian Quarterly* **17**: 172. . Retrieved June 7, 2010.
[37] Campbell, p. 120
[38] "A bridge too far along to be renamed" (http://special.registerguard.com/csp/cms/sites/web/sports/columnists/24574689-41/bridge-papÃ©-welch-commission-decision.csp). *The Register-Guard* (Eugene, Oregon): p. B1. March 18, 2010. . Retrieved July 29, 2011.
[39] Thornton, p. 125
[40] "Clackamas Indians" (http://www.nationalgeographic.com/lewisandclark/record_tribes_084_13_22.html). *National Geographic*. National Geographic Society. 1996. . Retrieved June 6, 2010.
[41] Snipp, p. 344
[42] Ruby and Brown, p. 4
[43] Ruby and Brown, p. 16
[44] Deur and Turner, p. 220
[45] "Willamette River Recreation Guide" (http://www.oregon.gov/OSMB/library/docs/WillametteRiverGuidePDF.pdf) (PDF). Oregon State Marine Board. 2007. pp. 1–3, 10. . Retrieved June 7, 2010.
[46] "Fur trade" (http://www.nwcouncil.org/history/FurTrade.asp). *Columbia River History*. Northwest Power and Conservation Council. 2010. . Retrieved June 6, 2010.
[47] Mackie, p. 115
[48] Engeman, p. 63
[49] Wilkes, pp. 341–374
[50] Meinig, p. 71
[51] "Introduction" (http://web.archive.org/web/20100529081949/http://www.isu.edu/~trinmich/Introduction.html). *The Oregon Trail*. Idaho State University. Archived from the original (http://www.isu.edu/~trinmich/Introduction.html) on May 29, 2010. .
[52] "The Route West" (http://web.archive.org/web/20100529081928/http://www.isu.edu/~trinmich/routewest.html). *The Oregon Trail*. Idaho State University. Archived from the original (http://www.isu.edu/~trinmich/routewest.html) on May 29, 2010. .
[53] "Where did the Oregon Trail Go? Reaching Oregon's Willamette Valley" (http://www.octa-trails.org/learn/people_places/articles_ot_by_here.php). *People & Places*. Oregon-California Trails Association. 2011. . Retrieved August 30, 2011.
[54] "Clackamas County" (http://bluebook.state.or.us/local/counties/counties03.htm). *Oregon Blue Book*. Oregon State Archives. 2011. . Retrieved February 24, 2011.
[55] Samson, p. 125
[56] Holman, pp. 96–97
[57] McArthur, p. 1022
[58] Gulick, pp. 28–29
[59] Gulick, p. 22
[60] Timmen, pp. 14, 17, 27
[61] Lewis, Alan (August 2006). "Conquering the Falls: The Willamette Falls Locks" (http://www.willamettefalls.org/HisLocks). *History of the Willamette Falls*. Willamette Falls Heritage Foundation. . Retrieved June 6, 2010.

[62] Balingit, Moriah; Lednicer, Lisa Grace (May 18, 2006). "Living on borrowed time" (http://www.willamettefalls.org/LocksNews). *The Oregonian* (Portland, Oregon). . Retrieved June 6, 2010.
[63] "A Chronology of Farming in Oregon" (http://www.opb.org/programs/oregonstory/farming/timeline.html). Oregon Public Broadcasting. 2000. . Retrieved June 6, 2010.
[64] Edwards and Schwantes, p. 61
[65] Edwards and Schwantes, p. 62
[66] Jette, Melinda. "Kalapuya Treaty of 1855" (http://www.oregonencyclopedia.org/entry/view/kalapuya_treaty/). *The Oregon Encyclopedia*. Portland State University. . Retrieved June 6, 2010.
[67] U.S. Coast and Geodetic Survey, p. 76
[68] Stenzel, pp. 37–39
[69] Willingham, William F. "U.S. Army Corps of Engineers" (http://www.oregonencyclopedia.org/entry/view/u_s__army_corps_of_engineers/). *The Oregon Encyclopedia*. Portland State University. . Retrieved September 14, 2011.
[70] "Sustainable Rivers: Willamette flow management" (http://www.nature.org/images/willamette-flow_factsheet_dec10.pdf) (PDF). *Conservation that works in Oregon*. The Nature Conservancy. . Retrieved June 6, 2010.
[71] "Willamette River basin" (http://www.nwr.noaa.gov/Salmon-Hydropower/Willamette-Basin/). National Marine Fisheries Service, Northwest Regional Office. November 6, 2009. . Retrieved September 8, 2011.
[72] Williams, p. 131
[73] "American Heritage Rivers News" (http://water.epa.gov/type/watersheds/named/heritage/10-5upda.cfm). U.S. Environmental Protection Agency. November 19, 2009. . Retrieved April 11, 2011.
[74] Bacon, Sheila (November 2003). "West side Big Pipe" (http://northwest.construction.com/features/archive/0311_Cover.asp). Northwest Construction. . Retrieved June 3, 2008.
[75] "East Side CSO Tunnel Project" (http://www.portlandonline.com/cso/index.cfm?c=31727). City of Portland. . Retrieved April 10, 2011.
[76] "A big pipe" (http://www.portlandonline.com/cso/index.cfm?a=180894&c=43918). *The Oregonian* (Portland, Oregon). December 11, 2007. . Retrieved June 3, 2008.
[77] "Willamette Falls Hydroelectric Project, Oregon" (http://www.lowimpacthydro.org/lihi-certificate-33-willamette-falls-hydroelectric-project-oregon.html). Low Impact Hydropower Institute. . Retrieved June 6, 2010.
[78] "Willamette basin reservoirs" (http://www.oregon.gov/OWRD/docs/WillametteReservoirs.pdf) (PDF). Oregon Water Resources Department; U.S. Army Corps of Engineers. . Retrieved September 15, 2011.
[79] "Willamette Valley locations" (http://www.nwp.usace.army.mil/locations/willamettevalley.asp). U.S. Army Corps of Engineers, Portland District. . Retrieved September 27, 2011.
[80] ACME Mapper. *USGS Topo Maps for United States* (http://mapper.acme.com/) (Map). Cartography by U.S. Geological Survey. . Retrieved June 7, 2010.
[81] Reinhardt, Bob. "Detroit Dam" (http://www.oregonencyclopedia.org/entry/view/detroit_dam/). *The Oregon Encyclopedia*. The Oregon Historical Society. . Retrieved June 7, 2010.
[82] "Willamette Basin" (http://www.salmonrecovery.gov/BiologicalOpinions/WillametteBasin.aspx). Federal Caucus. July 31, 2008. . Retrieved June 6, 2010.
[83] Smith, *et al.*, p. 287
[84] U.S. Geological Survey. "United States Geological Survey Topographic Map: Corvallis, Oregon, quadrangle" (http://www.topoquest.com/map.php?lat=44.56617&lon=-123.25734&datum=nad83&zoom=2&map=auto&coord=d&mode=zoomin&size=m). TopoQuest. . Retrieved May 31, 2010.
[85] Smith, *et al.*, p. 96
[86] U.S. Geological Survey. "United States Geological Survey Topographic Map: Oregon City, Oregon, quadrangle" (http://www.topoquest.com/map.php?lat=45.357222&lon=-122.607222&datum=nad83&zoom=4). TopoQuest. . Retrieved June 1, 2010.
[87] Smith, *et al.*, p. 78
[88] U.S. Geological Survey. "United States Geological Survey Topographic Map:Portland, Oregon, quadrangle" (http://www.topoquest.com/map.php?lat=45.50121&lon=-122.66454&datum=nad83&zoom=4). TopoQuest. . Retrieved June 1, 2010.
[89] Smith, *et al.*, p. 115
[90] Smith, *et al.*, p. 208
[91] Wortman and Norman, pp. 39–44
[92] Smith, *et al.*, p. 116
[93] Smith, *et al.*, p. 113
[94] U.S. Geological Survey. "United States Geological Survey Topographic Map:Linnton, Oregon, quadrangle" (http://www.topoquest.com/map.php?lat=45.58508&lon=-122.76477&datum=nad83&zoom=4). TopoQuest. . Retrieved June 5, 2010.
[95] Allen, Cain (2004). "The Great Flood of 1861" (http://www.ohs.org/education/oregonhistory/historical_records/dspDocument.cfm?doc_ID=06CA6808-E26A-5869-A052252B0B835B04). *Historical Records*. Oregon Historical Society. . Retrieved June 6, 2010.
[96] "News of the day: general news" (http://query.nytimes.com/mem/archive-free/pdf?res=F30F10F73D551A7493C3A9178CD85F428684F9) (PDF). *The New York Times* (New York, New York). July 1, 1866. . Retrieved July 2, 2011.

[97] "The floods and the storm" (http://news.google.com/newspapers?id=U8tVAAAAIBAJ&sjid=okANAAAAIBAJ&dq=willamette river flood -wiki&pg=4553,5634916). *The Philadelphia Record* (Philadelphia, Pennsylvania): p. 1. February 9, 1890. . Retrieved July 2, 2011.

[98] "Oregon's great floods" (http://query.nytimes.com/mem/archive-free/pdf?res=F50817F63F5F10738DDDAE0894DA405B8085F0D3) (PDF). *The New York Times* (New York, New York). February 7, 1890. . Retrieved July 2, 2011.

[99] "Floods wash out Oregon bridges" (http://news.google.com/newspapers?id=F_ZIAAAAIBAJ&sjid=QYkNAAAAIBAJ&dq=willamette river flood -wiki&pg=6691,1835508). *The Vancouver Sun* (Vancouver, British Columbia): p. 1. December 1, 1942. . Retrieved July 2, 2011.

[100] Associated Press (January 2, 1943). "Lower river areas hit by flood crest" (http://news.google.com/newspapers?id=W7BWAAAAIBAJ&sjid=aOgDAAAAIBAJ&dq=willamette river flood -wiki&pg=4605,2872262). *The Register-Guard* (Eugene, Oregon): p. 1. . Retrieved July 2, 2011.

[101] United Press International (January 1, 1943). "Flood peril wanes in Ohio Valley" (http://news.google.com/newspapers?id=t1wiAAAAIBAJ&sjid=epgFAAAAIBAJ&dq=willamette river flood -wiki&pg=3694,861829). *The Reading Eagle* (Reading, Pennsylvania): pp. 1, 18. . Retrieved July 2, 2011.

[102] Associated Press (November 28, 1946). "Willamette River is nearing crest of flood stage" (http://news.google.com/newspapers?id=RONXAAAAIBAJ&sjid=l_UDAAAAIBAJ&dq=willamette river flood -wiki&pg=7328,7090518). *The Spokesman-Review* (Spokane, Washington): p. 2. . Retrieved July 2, 2011.

[103] Associated Press (December 14, 1948). "River flood waters reach Oregon capital" (http://news.google.com/newspapers?id=dsBkAAAAIBAJ&sjid=AHUNAAAAIBAJ&dq=willamette river flood -wiki&pg=1317,5881058). *The Telegraph-Herald* (Dubuque, Iowa): p. 16. . Retrieved July 2, 2011.

[104] Gregory, L.; Jett, S.; Wildman, R. "Floods and Floodplains in the Willamette River" (http://www.fsl.orst.edu/pnwerc/wrb/Atlas_web_compressed/3.Water_Resources/3e.flood&fema_web.pdf) (PDF). *Federal Emergency Management Agency*. Corvallis Forestry Research Community. . Retrieved June 6, 2010.

[105] "Portland threatened by rampaging river" (http://news.google.com/newspapers?id=OGEhAAAAIBAJ&sjid=1oYFAAAAIBAJ&dq=portland threatened by rampaging river&pg=1568,4038947). *The Tri-City Herald* (Kennewick, Washington): p. 20. December 21, 1964. . Retrieved June 29, 2011.

[106] "Floods spoil holiday in Oregon, California" (http://news.google.com/newspapers?id=Nq4RAAAAIBAJ&sjid=wOgDAAAAIBAJ&dq=willamette flood -wiki&pg=7350,2520105). *The Spokesman-Review* (Spokane, Washington): p. 2. December 25, 1964. . Retrieved June 29, 2011.

[107] "Portland gets reprieve from a big flood threat" (http://news.google.com/newspapers?id=-JUfAAAAIBAJ&sjid=S9QEAAAAIBAJ&dq=portland gets reprieve from a big flood threat&pg=1976,5832349). *The Nevada Daily Mail* (Nevada, Missouri): p. 1. December 27, 1964. . Retrieved June 29, 2011.

[108] United Press International (January 30, 1965). "1000 flee flood in Northwest; nation shivers" (http://news.google.com/newspapers?id=iF4bAAAAIBAJ&sjid=O08EAAAAIBAJ&dq=willamette river flood -wiki&pg=6917,4443507). *The Pittsburgh Press* (Pittsburgh, Pennsylvania): p. 1. . Retrieved July 2, 2011.

[109] Associated Press (February 6, 1965). "Eastern deep freeze eases, but rain brings flood fears to West" (http://news.google.com/newspapers?id=GsZGAAAAIBAJ&sjid=0_MMAAAAIBAJ&dq=willamette river flood -wiki&pg=6981,2178938). *The Lewiston Daily Sun* (Lewiston, Maine): p. 1. . Retrieved July 2, 2011.

[110] Associated Press (February 10, 1996). "Swollen river spares downtown Portland" (http://news.google.com/newspapers?id=X_BHAAAAIBAJ&sjid=Ie0DAAAAIBAJ&dq=the river crested at one town after another like a meal moving through a boa constrictor&pg=6510,5004315). *The Deseret News* (Salt Lake City, Utah): p. A4. . Retrieved June 25, 2011.

[111] Taylor, George H. (November 1996). "The Great Flood of 1996: its causes and a comparison to other climate events" (http://ir.library.oregonstate.edu/xmlui/bitstream/handle/1957/6134/SR no.970_ocr.pdf?sequence=1) (PDF). Oregon State University. p. 2. . Retrieved August 3, 2011.

[112] Barnard, Jeff (February 25, 1996). "Flood spreads in economy like muddy water" (http://nl.newsbank.com/nl-search/we/Archives?p_product=RM&p_theme=rm&p_action=search&p_maxdocs=200&p_topdoc=1&p_text_direct-0=0EB4E49FC4CAF785&p_field_direct-0=document_id&p_perpage=10&p_sort=YMD_date:D&s_trackval=GooglePM). *The Rocky Mountain News* (Denver, Colorado). . Retrieved June 26, 2011.

[113] "Flood wall comes down today in Portland" (http://news.google.com/newspapers?id=ZEdWAAAAIBAJ&sjid=6-oDAAAAIBAJ&dq=willamette river flood 1996 -wiki&pg=5507,82622). *The Register-Guard* (Eugene, Oregon): p. 3C. April 1, 1996. . Retrieved June 26, 2011.

[114] Associated Press (October 14, 1996). "Flood wall readied for next time" (http://news.google.com/newspapers?id=MUxWAAAAIBAJ&sjid=HOsDAAAAIBAJ&dq=willamette steel plates -wiki&pg=1859,3852913). *The Register-Guard* (Eugene, Oregon): p. 5C. . Retrieved June 27, 2011.

[115] Williams, p. 132

[116] The Los Angeles Times; The Washington Post (April 17, 1973). "Oregon sets pace in improving environment" (http://news.google.com/newspapers?id=8pscAAAAIBAJ&sjid=sWYEAAAAIBAJ&dq=tom mccall willamette river&pg=7577,366713). *The Sarasota Herald-Tribune* (Sarasota, Florida): p. 20A. . Retrieved July 17, 2011.

[117] "Portland Harbor Cleanup Sites" (http://www.deq.state.or.us/lq/cu/nwr/portlandharbor/docs/PortlandHarborMap.pdf) (PDF). Oregon Department of Environmental Quality. February 17, 2009. . Retrieved April 11, 2011.

[118] Lee van der Voo (July 3, 2007). "History of a Superfund cleanup bid" (http://portlandtribune.com/news/story.php?story_id=118341365933266700). *Portland Tribune* (Portland, Oregon). . Retrieved February 12, 2008.

[119] "Portland Harbor Superfund Site" (http://web.archive.org/web/20061216212032/http://www.deq.state.or.us/nwr/PortlandHarbor/ph.htm). Oregon Department of Environmental Quality. Archived from the original (http://www.deq.state.or.us/nwr/PortlandHarbor/ph.htm) on December 16, 2006. . Retrieved December 16, 2006.

[120] "Region 10 Cleanup: Portland Harbor" (http://yosemite.epa.gov/r10/cleanup.nsf/sites/ptldharbor). U.S. Environmental Protection Agency. . Retrieved February 12, 2008.

[121] "The Willamette River: a renaissance in the making" (http://www.americanrivers.org/assets/pdfs/reports-and-publications/chapter_6e1f9.pdf) (PDF). American Rivers. p. 140. . Retrieved July 11, 2011.

[122] "Processes Controlling Dissolved Oxygen and pH in the Upper Willamette River Basin" (http://or.water.usgs.gov/pubs_dir/Pdf/95-4205.pdf) (PDF). U.S. Geological Survey. 1994. . Retrieved February 12, 2008.

[123] "Drinking the Willamette – the power of public education" (http://www.pnws-awwa.org/Files/Drinking the Willamette.pdf) (PDF). Pacific Northwest Section of the American Water Works Association 2008 Annual Conference. May 1, 2008. p. 4. . Retrieved July 11, 2011.

[124] Oregon Water Quality Index, p. 6

[125] Oregon Water Quality Index, pp. 7–10

[126] Benke, *et al.*, p. 619

[127] Benke, *et al.*, pp. 620–21

[128] "Willamette Valley projects" (http://www.ohjv.org/projects/willamette.html). Oregon Habitat Joint Venture. November 14, 2001. . Retrieved July 11, 2011.

[129] "Willamette Valley National Wildlife Refuge Complex" (http://www.fws.gov/WillametteValley/complex/). U.S. Fish and Wildlife Service. July 18, 2010. . Retrieved July 11, 2011.

[130] Zaveri, Mihir (June 17, 2011). "In Oregon's Willamette Valley, the race to save the endangered Fender's blue butterfly". *The Oregonian* (Portland, Oregon).

[131] Benke, *et al.*, p. 621

References

Works cited

- — "Oregon Water Quality Index Summary Report, Water Years 1998–2007" (http://www.deq.state.or.us/lab/wqm/docs/09-LAB-008.pdf), May 2008. Oregon Department of Environmental Quality. Retrieved April 10, 2011.
- Benke, Arthur C., ed., and Cushing, Colbert E., ed.; Stanford, Jack A.; Gregory, Stanley V.; Hauer, Richard F.; Snyder, Eric B. (2005). "Chapter 13: Columbia River Basin" in *Rivers of North America*. Burlington, Massachusetts: Elsevier Academic Press. ISBN 0-12-088253-1. OCLC 59003378.
- Bright, William (2004). *Native American Placenames of the United States*. Norman, Oklahoma: University of Oklahoma Press. ISBN 0-8061-3576-X. OCLC 53019644.
- Campbell, Lyle (1997). *American Indian Languages*. New York, New York: Oxford University Press. ISBN 0-19-509427-1. OCLC 247389024.
- Deur, Douglas; Turner, Nancy J. (2005). *Keeping it Living: Traditions of Plant Use and Cultivation on the Northwest Coast of North America*. Seattle: University of Washington Press. ISBN 0-77481-267-2. OCLC 57316620.
- Edwards, G. Thomas; Schwantes, Carlos A. (1986). *Experiences in a Promised Land: Essays in Pacific Northwest History*. Seattle: University of Washington Press. ISBN 0-29596-328-X. OCLC 12805717.
- Engeman, Richard H. (2009). *The Oregon Companion: A Historical Gazetteer of the Useful, the Curious, and the Arcane*. Portland, Oregon: Timber Press. ISBN 0-88192-899-2. OCLC 236142647.
- Gulick, Bill (2004). *Steamboats on Northwest Rivers*. Caldwell, Idaho: Caxton Press. ISBN 0-87004-438-9. OCLC 53000219.
- Holman, Frederick Van Voorhies (1907). *Dr. John McLoughlin, the Father of Oregon*. Cleveland, Ohio: A. H. Clark Company. ISBN 1-13113-770-1. OCLC 365933029.
- Laenen, Antonius; Dunnette, David A. (1997). *River Quality: Dynamics and Restoration*. Boca Raton, Florida: CRC Press. ISBN 1-56670-138-4. OCLC 34885999.

- Loy, Willam G.; Allan, Stuart Allan; Buckley, Aileen R.; Meecham, James E. (2001) [1976]. *Atlas of Oregon* (2nd ed.). Eugene, Oregon: University of Oregon Press. ISBN 0-87114-101-9. OCLC 228223337.
- Mackie, Richard Somerset (1997). *Trading Beyond the Mountains: the British Fur Trade on the Pacific, 1793–1843*. Vancouver: UBC Press. ISBN 0-77480-613-3. OCLC 82135549.
- McArthur, Lewis A.; McArthur, Lewis L. (2003) [1928]. *Oregon Geographic Names* (7th ed.). Portland, Oregon: Oregon Historical Society Press. ISBN 0-87595-277-1. OCLC 53075956.
- Meinig, D.W. (1998). *The Shaping of America: A Geographical Perspective on 500 Years of History, Volume 3: Transcontinental America, 1850–1915*. New Haven, Connecticut: Yale University Press. ISBN 0-300-08290-8. OCLC 60197763.
- Orr, Elizabeth L.; Orr, William N. (1999). *Geology of Oregon* (5th ed.). Dubuque, Iowa: Kendall/Hunt Publishing Company. ISBN 0-7872-6608-6. OCLC 42944922.
- Ruby, Robert H.; Brown, John A. (1976). *The Chinook Indians: Traders of the Lower Columbia River*. Civilization of the American Indian. Norman, Oklahoma: University of Oklahoma Press. ISBN 0-80612-107-6. OCLC 1958350.
- Samson, Karl (2010). *Frommer's Oregon*. Hoboken, New Jersey: Wiley-Blackwell. ISBN 0-47053-771-X. OCLC 671277501.
- Smith, Dwight A.; Norman, James B.; Dykman, Pieter T. (1989) [1986]. *Historic Highway Bridges of Oregon* (2nd ed.). Portland, Oregon: Oregon Historical Society Press. ISBN 0-87595-205-4. OCLC 20362124.
- Snipp, C. Matthew (1989). *American Indians: The First of This Land*. New York, New York: Russell Sage Foundation. ISBN 0-87154-822-4. OCLC 19670797.
- Stenzel, Franz (1972). *Cleveland Rockwell: Scientist and Artist, 1837–1907*. Portland, Oregon: Oregon Historical Society. ISBN 0-87595-037-X. OCLC 333762.
- Thornton, Russell (1990). *American Indian Holocaust and Survival: A Population History Since 1492*. Civilization of the American Indian. Norman, Oklahoma: University of Oklahoma Press. ISBN 0-80612-220-X. OCLC 263150623.
- Timmen, Fritz (1973). *Blow for the Landing*. Atglen, Pennsylvania: Schiffer. OCLC 38286714.
- U.S. Coast and Geodetic Survey (1887). *Report of the Superintendent of the U.S. Coast and Geodetic Survey Showing the Progress of the Work during the Fiscal Year Ending with June 1886*. Washington, D.C.: Government Printing Office. OCLC 28937852.
- Wilkes, Charles (1845). *Narrative of the United States' Exploring Expedition, During the Years 1838, 1839, 1840, 1841, 1842: Condensed and abridged* (http://books.google.com/?id=o3AFAAAAQAAJ&printsec=frontcover#v=onepage&q&f=false). London: Whittaker and Co. OCLC 602918838.
- Williams, Travis (2009). *The Willamette River Field Guide*. Portland, Oregon: Timber Press. ISBN 0-88192-866-6. OCLC 243601804.
- Wortman, Sharon Wood; Wortman, Ed.; Norman, James B. (2006). *The Portland Bridge Book* (3rd ed.). Portland, Oregon: Urban Adventure Press. ISBN 0-9787365-1-6. OCLC 76928209.

External links

- City of Portland River Renaissance Strategy (http://www.portlandonline.com/auditor/index.cfm?a=72561&c=35191)
- Guiding the Willamette (http://opb.org/thinkoutloud/shows/guiding-willamette/), *Think Out Loud* radio program on Oregon Public Broadcasting, April 1, 2009
- Portland Harbor DEQ cleanup (http://www.deq.state.or.us/lq/cu/nwr/PortlandHarbor/index.htm) – Oregon State Department of Environmental Quality
- Willamette Riverkeeper (http://www.willamette-riverkeeper.org) – Non-profit environmental organization

List of United States cities by population

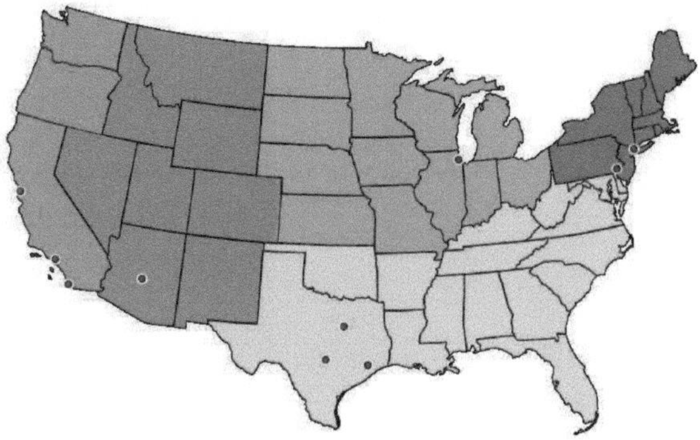

Ten most populous cities in the United States

(2) Los Angeles

(10) San Jose

(8) San Diego

(6) Phoenix

(3) Chicago

(1) New York

(4) Houston

(7) San Antonio

(9) Dallas

(5) Philadelphia

The following is a **list of the most populous incorporated places in the United States**. As defined by the United States Census Bureau, an "incorporated place" includes a variety of designations, including a city, town, village, borough, and municipality.[a] Some census-designated places may also be included in the Census Bureau's listing of incorporated places.[b][c] Consolidated city-counties represent a distinct type of government that includes the entire population of a county, or county equivalent. Some consolidated city-counties, however, include multiple incorporated places. This list presents only that portion (or "balance") of such consolidated city-counties that are not a part of another incorporated place.

Note that this list refers only to the population of individual municipalities within their defined limits, which does not include other municipalities or unincorporated suburban areas within urban agglomerations. A different ranking is evident when considering U.S. metropolitan area populations.

Incorporated places over 100,000 population

The following table lists the incorporated places in the United States with a population on April 1, 2010, of more than 100,000 as counted in the 2010 United States Census.[1] Land area and population density are also provided to the nearest tenth to give a general idea of how densely packed each city is. Since the U.S. Census Bureau typically does not rank the cities of U.S. territories along with those of the fifty states and the District of Columbia, the cities of Puerto Rico with populations greater than 100,000 are listed separately below. Capital cities are **bolded**.

Rank	City	State	Population (2010)	Land Area (square miles, 2010 data)	Population Density (per square mile of land area, 2010 data)
1	New York	New York	8175133	302.6	27016.3
2	Los Angeles	California	3792621	468.7	8091.8
3	Chicago	Illinois	2695598	227.6	11843.6
4	Houston	Texas	2099451	599.6	3501.4
5	Philadelphia	Pennsylvania	1526006	134.1	11379.6
6	**Phoenix**	Arizona	1445632	516.7	2797.8
7	San Antonio	Texas	1327407	460.9	2880.0
8	San Diego	California	1307402	325.2	4020.3
9	Dallas	Texas	1197816	340.5	3517.8
10	San Jose	California	945942	176.5	5359.4
11	Jacksonville[g]	Florida	821784	747.0	1100.1
12	**Indianapolis**[f]	Indiana	820445	361.4	2270.2
13	San Francisco	California	805235	46.9	17169.2
14	**Austin**	Texas	790390	297.9	2653.2
15	**Columbus**	Ohio	787033	217.2	3623.5
16	Fort Worth	Texas	741206	339.8	2181.3
17	Charlotte	North Carolina	731424	297.7	2456.9
18	Detroit	Michigan	713777	138.8	5142.5
19	El Paso	Texas	649121	255.2	2543.6
20	Memphis	Tennessee	646889	315.1	2053.0
21	Baltimore	Maryland	620961	80.9	7675.7
22	**Boston**	Massachusetts	617594	48.3	12786.6
23	Seattle	Washington	608660	83.9	7254.6
24	**Washington**	District of Columbia	601723	61.0	9864.3
25	**Nashville** [f]	Tennessee	601222	475.1	1265.5
26	**Denver**	Colorado	600158	153.0	3922.6
27	Louisville [f]	Kentucky	597337	325.2	1836.8
28	Milwaukee	Wisconsin	594833	96.1	6189.7
29	Portland	Oregon	583776	134.3	4346.8
30	Las Vegas	Nevada	583756	135.8	4298.6
31	**Oklahoma City**	Oklahoma	579999	606.4	956.5
32	Albuquerque	New Mexico	545852	187.7	2908.1
33	Tucson	Arizona	520116	226.7	2294.3
34	Fresno	California	494665	112.0	4416.7
35	**Sacramento**	California	466488	97.9	4764.9
36	Long Beach	California	462257	50.3	9190.0

37	Kansas City	Missouri	459787	315.0	1459.6
38	Mesa	Arizona	439041	136.5	3216.4
39	Virginia Beach [e]	Virginia	437994	249.0	1759.0
40	**Atlanta**	Georgia	420003	133.2	3153.2
41	Colorado Springs	Colorado	416427	194.5	2141.0
42	Omaha	Nebraska	408958	127.1	3217.6
43	**Raleigh**	North Carolina	403892	142.9	2826.4
44	Miami	Florida	399457	35.9	11126.9
45	Cleveland	Ohio	396815	77.7	5107.0
46	Tulsa	Oklahoma	391906	196.8	1991.4
47	Oakland	California	390724	55.8	7002.2
48	Minneapolis	Minnesota	382578	54.0	7084.8
49	Wichita	Kansas	382368	159.3	2400.3
50	Arlington	Texas	365438	95.9	3810.6
51	Bakersfield	California	347483	142.2	2443.6
52	New Orleans	Louisiana	343829	169.4	2029.7
53	**Honolulu** [b]	Hawaii	337256	60.5	5574.5
54	Anaheim	California	336265	49.8	6752.3
55	Tampa	Florida	335709	113.4	2960.4
56	Aurora	Colorado	325078	154.7	2101.3
57	Santa Ana	California	324528	27.3	11887.5
58	Saint Louis [d]	Missouri	319294	61.9	5158.2
59	Pittsburgh	Pennsylvania	305704	55.4	5518.1
60	Corpus Christi	Texas	305215	160.6	1900.5
61	Riverside	California	303871	81.1	3746.9
62	Cincinnati	Ohio	296943	77.9	3811.8
63	Lexington	Kentucky	295803	283.6	1043.0
64	Anchorage	Alaska	291826	1704.7	171.2
65	Stockton	California	291707	61.7	4727.8
66	Toledo	Ohio	287208	80.7	3559.0
67	**Saint Paul**	Minnesota	285068	52.0	5482.1
68	Newark	New Jersey	277140	24.2	11452.1
69	Greensboro	North Carolina	269666	126.5	2131.7
70	Buffalo	New York	261310	40.4	6468.1
71	Plano	Texas	259841	71.6	3629.1
72	**Lincoln**	Nebraska	258379	89.1	2899.9
73	Henderson	Nevada	257729	107.7	2393.0
74	Fort Wayne	Indiana	253691	110.6	2293.8
75	Jersey City	New Jersey	247597	14.8	16729.5

76	Saint Petersburg	Florida	244769	61.7	3967.1
77	Chula Vista	California	243916	49.6	4917.7
78	Norfolk [e]	Virginia	242803	54.1	4488.0
79	Orlando	Florida	238300	102.4	2327.1
80	Chandler	Arizona	236123	64.4	3666.5
81	Laredo	Texas	236091	88.9	2655.7
82	**Madison**	Wisconsin	233209	76.8	3036.6
83	Winston-Salem	North Carolina	229617	132.4	1734.3
84	Lubbock	Texas	229573	122.4	1875.6
85	**Baton Rouge**	Louisiana	229493	76.9	2984.3
86	Durham	North Carolina	228330	107.4	2126.0
87	Garland	Texas	226876	57.1	3973.3
88	Glendale	Arizona	226721	60.0	3778.7
89	Reno	Nevada	225221	103.0	2186.6
90	Hialeah	Florida	224669	21.5	10449.7
91	Chesapeake [e]	Virginia	222209	340.8	652.0
92	Scottsdale	Arizona	217385	183.9	1182.1
93	North Las Vegas	Nevada	216961	101.3	2141.8
94	Irving	Texas	216290	67.0	3228.2
95	Fremont	California	214089	77.5	2762.4
96	Irvine	California	212375	66.1	3212.9
97	Birmingham	Alabama	212237	146.1	1452.7
98	Rochester	New York	210565	35.8	5881.7
99	San Bernardino	California	209924	59.2	3546.0
100	Spokane	Washington	208916	59.2	3529.0
101	Gilbert	Arizona	208453	68.0	3065.5
102	Arlington [c]	Virginia	207627	26.0	7985.7
103	**Montgomery**	Alabama	205764	159.6	1289.2
104	**Boise**	Idaho	205671	79.4	2590.3
105	**Richmond** [e]	Virginia	204214	59.8	3414.9
106	**Des Moines**	Iowa	203433	80.9	2514.6
107	Modesto	California	201165	36.9	5451.6
108	Fayetteville	North Carolina	200654	145.8	1376.2
109	Shreveport	Louisiana	199311	105.4	1891.0
110	Akron	Ohio	199110	62.0	3211.5
111	Tacoma	Washington	198397	49.7	3991.9
112=	Aurora	Illinois	197899	44.9	4407.6
112=	Oxnard	California	197899	26.9	7356.8

114	Fontana	California	196069	42.4	4624.3
115	Yonkers	New York	195976	18.0	10887.6
116	Augusta [f]	Georgia	195844	302.5	647.4
117	Mobile	Alabama	195111	139.1	1402.7
118	**Little Rock**	Arkansas	193524	119.2	1623.5
119	Moreno Valley	California	193365	51.3	3769.3
120	Glendale	California	191719	30.5	6285.9
121	Amarillo	Texas	190695	99.5	1916.5
122	Huntington Beach	California	189992	26.7	7115.8
123	Columbus	Georgia	189885	216.4	877.5
124	Grand Rapids	Michigan	188040	44.4	4235.1
125	**Salt Lake City**	Utah	186440	111.1	1678.1
126	**Tallahassee**	Florida	181376	100.2	1810.1
127	Worcester	Massachusetts	181045	37.4	4840.8
128	Newport News [e]	Virginia	180719	68.7	2630.6
129	Huntsville	Alabama	180105	209.1	861.3
130	Knoxville	Tennessee	178874	98.5	1816.0
131	**Providence**	Rhode Island	178042	18.4	9676.2
132	Santa Clarita	California	176320	52.7	3345.7
133	Grand Prairie	Texas	175396	72.1	2432.7
134	Brownsville	Texas	175023	132.3	1322.9
135	**Jackson**	Mississippi	173514	111.0	1563.2
136	Overland Park	Kansas	173372	74.8	2317.8
137	Garden Grove	California	170883	17.9	9546.5
138	Santa Rosa	California	167815	41.3	4063.3
139	Chattanooga	Tennessee	167674	137.2	1222.1
140	Oceanside	California	167086	41.2	4055.5
141	Fort Lauderdale	Florida	165521	34.8	4756.4
142	Rancho Cucamonga	California	165269	39.9	4142.1
143	Port Saint Lucie	Florida	164603	114.0	1443.9
144	Ontario	California	163924	49.9	3285.1
145	Vancouver	Washington	161791	46.5	3479.4
146	Tempe	Arizona	161719	39.9	4053.1
147	Springfield	Missouri	159498	81.7	1952.2
148	Lancaster	California	156633	94.3	1661.0
149	Eugene	Oregon	156185	43.7	3574.0
150	Pembroke Pines	Florida	154750	33.1	4675.2
151	**Salem**	Oregon	154637	47.9	3228.3
152	Cape Coral	Florida	154305	105.7	1459.8

153	Peoria	Arizona	154065	174.4	883.4
154	Sioux Falls	South Dakota	153888	73.0	2108.1
155	Springfield	Massachusetts	153060	31.9	4798.1
156	Elk Grove	California	153015	42.2	3625.9
157	Rockford	Illinois	152871	61.1	2502.0
158	Palmdale	California	152750	106.0	1441.0
159	Corona	California	152374	38.8	3927.2
160	Salinas	California	150441	23.2	6484.5
161	Pomona	California	149058	23.0	6480.8
162	Pasadena	Texas	149043	42.8	3482.3
163	Joliet	Illinois	147433	62.1	2374.1
164	Paterson	New Jersey	146199	8.4	17404.6
165	Kansas City	Kansas	145786	124.8	1168.2
166	Torrance	California	145438	20.5	7094.5
167	Syracuse	New York	145170	25.0	5806.8
168	Bridgeport	Connecticut	144229	16.0	9014.3
169	Hayward	California	144186	45.3	3182.9
170	Fort Collins	Colorado	143986	54.3	2651.7
171	Escondido	California	143911	36.8	3910.6
172	Lakewood	Colorado	142980	42.9	3332.9
173	Naperville	Illinois	141853	38.8	3656.0
174	Dayton	Ohio	141527	55.7	2540.9
175	Hollywood	Florida	140768	27.4	5137.5
176	Sunnyvale	California	140081	22.0	6367.3
177	Alexandria [e]	Virginia	139966	15.0	9331.1
178	Mesquite	Texas	139824	46.0	3039.7
179	Hampton [e]	Virginia	137436	51.4	2673.9
180	Pasadena	California	137122	23.0	5961.8
181	Orange	California	136416	24.8	5500.6
182	Savannah	Georgia	136286	103.2	1320.6
183	Cary	North Carolina	135234	54.3	2490.5
184	Fullerton	California	135161	22.4	6034.0
185	Warren	Michigan	134056	34.4	3897.0
186	Clarksville	Tennessee	132929	97.6	1362.0
187	McKinney	Texas	131117	62.2	2108.0
188	McAllen	Texas	129877	48.3	2689.0
189	New Haven	Connecticut	129779	18.7	6940.1
190	Sterling Heights	Michigan	129699	36.5	3553.4
191	West Valley City	Utah	129480	35.6	3637.1

192	**Columbia**	South Carolina	129272	132.2	977.9
193	Killeen	Texas	127921	53.6	2386.6
194	**Topeka**	Kansas	127473	60.2	2117.5
195	Thousand Oaks	California	126683	55.0	2303.3
196	Cedar Rapids	Iowa	126326	70.8	1784.3
197	Olathe	Kansas	125872	59.7	2108.4
198	Elizabeth	New Jersey	124969	12.3	10160.1
199	Waco	Texas	124805	89.0	1402.3
200	**Hartford**	Connecticut	124775	17.4	7171.0
201	Visalia	California	124442	36.2	3437.6
202	Gainesville	Florida	124354	61.3	2028.6
203	Simi Valley	California	124237	41.5	2993.7
204	Stamford	Connecticut	122643	37.6	3261.8
205	Bellevue	Washington	122363	32.0	3823.8
206	Concord	California	122067	30.5	4002.2
207	Miramar	Florida	122041	29.5	4137.0
208	Coral Springs	Florida	121096	23.8	5088.1
209	Lafayette	Louisiana	120623	49.2	2451.7
210	Charleston	South Carolina	120083	109.0	1101.7
211	Carrollton	Texas	119097	36.3	3280.9
212	Roseville	California	118788	36.2	3281.4
213	Thornton	Colorado	118772	34.8	3413.0
214	Beaumont	Texas	118296	82.8	1428.7
215	Allentown	Pennsylvania	118032	17.5	6744.7
216	Surprise	Arizona	117517	105.7	1111.8
217	Evansville	Indiana	117429	44.2	2656.8
218	Abilene	Texas	117063	106.8	1096.1
219	Frisco	Texas	116989	61.8	1893.0
220	Independence	Missouri	116830	77.6	1505.5
221	Santa Clara	California	116468	18.4	6329.8
222	**Springfield**	Illinois	116250	59.5	1953.8
223	Vallejo	California	115942	30.7	3776.6
224	Victorville	California	115903	73.2	1583.4
225	Athens [f]	Georgia	115452	116.4	991.9
226	Peoria	Illinois	115007	48.0	2396.0
227	**Lansing**	Michigan	114297	36.0	3174.9
228	Ann Arbor	Michigan	113934	27.8	4098.3
229	El Monte	California	113475	9.6	11820.3
230	Denton	Texas	113383	88.0	1288.4

231	Berkeley	California	112580	10.5	10721.9
232	Provo	Utah	112488	41.7	2697.6
233	Downey	California	111772	12.4	9013.9
234	Midland	Texas	111147	72.1	1541.6
235	Norman	Oklahoma	110925	178.8	620.4
236	Waterbury	Connecticut	110366	28.5	3872.5
237	Costa Mesa	California	109960	15.7	7003.8
238	Inglewood	California	109673	9.1	12052.0
239	Manchester	New Hampshire	109565	33.1	3310.1
240	Murfreesboro	Tennessee	108755	55.3	1966.6
241	Columbia	Missouri	108500	63.1	1719.5
242	Elgin	Illinois	108188	37.2	2908.3
243	Clearwater	Florida	107685	25.6	4206.4
244	Miami Gardens	Florida	107167	18.2	5888.3
245	Rochester	Minnesota	106769	54.6	1955.5
246	Pueblo	Colorado	106595	53.6	1988.7
247	Lowell	Massachusetts	106519	13.6	7832.3
248	Wilmington	North Carolina	106476	51.5	2067.5
249=	Arvada	Colorado	106433	35.1	3032.3
249=	San Buenaventura (Ventura)	California	106433	21.7	4904.7
251	Westminster	Colorado	106114	31.6	3358.0
252	West Covina	California	106098	16.0	6631.1
253	Gresham	Oregon	105594	23.2	4551.5
254=	Fargo	North Dakota	105549	48.8	2162.9
254=	Norwalk	California	105549	9.7	10881.3
256	Carlsbad	California	105328	37.7	2793.8
257	Fairfield	California	105321	37.4	2816.1
258	Cambridge	Massachusetts	105162	6.4	16431.6
259	Wichita Falls	Texas	104553	72.1	1450.1
260	High Point	North Carolina	104371	53.8	1940.0
261	Billings	Montana	104170	43.4	2400.2
262	Green Bay	Wisconsin	104057	45.5	2287.0
263	West Jordan	Utah	103712	32.5	3191.1
264	Richmond	California	103701	30.1	3445.2
265	Murrieta	California	103466	33.6	3079.3
266	Burbank	California	103340	17.3	5973.4
267	Palm Bay	Florida	103190	65.7	1570.6
268	Everett	Washington	103019	33.4	3084.4
269	Flint	Michigan	102434	33.4	3066.9

270	Antioch	California	102372	28.3	3617.4
271	Erie	Pennsylvania	101786	19.1	5329.1
272	South Bend	Indiana	101168	41.5	2437.8
273	Daly City	California	101123	7.7	13132.9
274	Centennial	Colorado	100377	28.7	3497.5
275	Temecula	California	100097	30.2	3314.5

Gallery

1 - New York, New York

2 - Los Angeles, California

3 - Chicago, Illinois

4 - Houston, Texas

5 - Philadelphia, Pennsylvania

6 - Phoenix, Arizona

7 - San Antonio, Texas

8 – San Diego, California

9 – Dallas, Texas

10 – San Jose, California

11 – Jacksonville, Florida

12 – Indianapolis, Indiana

13 – San Francisco, California

14 – Austin, Texas

15 – Columbus, Ohio

16 – Fort Worth, Texas

17 – Charlotte, North Carolina

18 – Detroit, Michigan

19 – El Paso, Texas

20 – Memphis, Tennessee

21 – Baltimore, Maryland

22 – Boston, Massachusetts

23 – Seattle, Washington

24 – Washington, D.C.

25 – Nashville, Tennessee

26 – Denver, Colorado

27 – Louisville, Kentucky

28 – Milwaukee, Wisconsin

29 – Portland, Oregon

30 – Las Vegas, Nevada

31 – Oklahoma City, Oklahoma

32 – Albuquerque, New Mexico

33 - Tucson, Arizona

34 – Fresno, California

35 – Sacramento, California

36 – Long Beach, California

37 – Kansas City, Missouri

38 – Mesa, Arizona

39 – Virginia Beach, Virginia

40 – Atlanta, Georgia

41 – Colorado Springs, Colorado

42 – Omaha, Nebraska

43 – Raleigh, North Carolina

44 – Miami, Florida

45 – Cleveland, Ohio

46 – Tulsa, Oklahoma

47 – Oakland, California

48 – Minneapolis, Minnesota

49 - Wichita, Kansas

50 – Arlington, Texas

Puerto Rico

The following table lists the incorporated municipalities (*municipios*) of Puerto Rico with a population greater than 100,000 according to the 2010 United States Census,[2] and the rankings they would have if included in the above table.

(46) – San Juan, Puerto Rico

Rank	City	Population (2009)
46	**San Juan**	395326
103	Bayamón	208116
134	Carolina	176762
144	Ponce	166327
176	Caguas	142893

See also

- Largest cities in the United States by population by decade
- List of cities proper by population (most populous cities in the World)
- List of lists of settlements in the United States
- List of United States cities by area
- List of United States cities by population density
- List of United States urban areas
- List of U.S. states and territories by population
- Table of United States Combined Statistical Areas (CSA)
- Table of United States Core Based Statistical Areas (CBSA)
- Table of United States Metropolitan Statistical Areas (MSA)
- Table of United States Micropolitan Statistical Areas (μSA)

Notes

a. ^ Towns in New England, while incorporated on a level similar to cities in other states, are considered minor civil divisions by the Census Bureau and are not included in its list of incorporated places. A detailed discussion on this subject can be found at New England town#Census treatment of the New England town system.

b. ^ *a b* Hawaii has no incorporated municipalities other than the City and County of Honolulu, which actually includes the entire island of Oahu. In accordance with Hawaiian law, the U.S. Census Bureau defines the state's "cities" and "towns" as census-designated places (CDPs). The Census Bureau defines Honolulu CDP as the portion of the City and County that is coextensive with the Judicial District of Honolulu. Honolulu CDP is what is generally thought of as the "city" of Honolulu, and its population is used here and in other population comparisons. If the entire population of the City and County were included, Honolulu would place 10th on the list above with a 2010 estimated population of 953,207.[3]

c. ^ *a b* Arlington CDP, Virginia, is coextensive with Arlington County. Virginia law forbids the creation of cities in counties with over 1,000 people per square mile, which Arlington County exceeds. Thus, the entire county is considered by the Census Bureau as one CDP. While not incorporated as a municipality, Arlington is included in the Census Bureau's annual list of incorporated places and is retained in the above list.

d. ^ *a b* Baltimore and St. Louis are independent cities that are not a part of any particular county.

e. ^ *a b c d* Cities in Virginia, except for Arlington, are all independent cities that are not a part of any particular county.

f. ^ *a b c d e* Indianapolis, Indiana; Louisville, Kentucky; Nashville, Tennessee; Augusta, Georgia; and Athens, Georgia are consolidated city–counties. The population given is for the entire city, excluding other incorporated places lying within the county limits. (See Indianapolis (balance), Indiana; Louisville – Jefferson County (balance), Kentucky; and Nashville-Davidson (balance), Tennessee). As of 2009, the consolidated populations of these city-counties *including* other incorporated places were, respectively: 890,879;[4] 721,594;[5] 635,710;[6] 199,768;[7] and 116,342.[7]

g. ^ Jacksonville, Florida is a consolidated city-county with Duval County. The population given is for the city excluding other incorporated places within Duval County. As of 2010 the consolidated population including other incorporated populations was 864,263.

Distributions

For cities with populations of 100,000 or more the following distributions hold. Smaller incorporated places are not included.

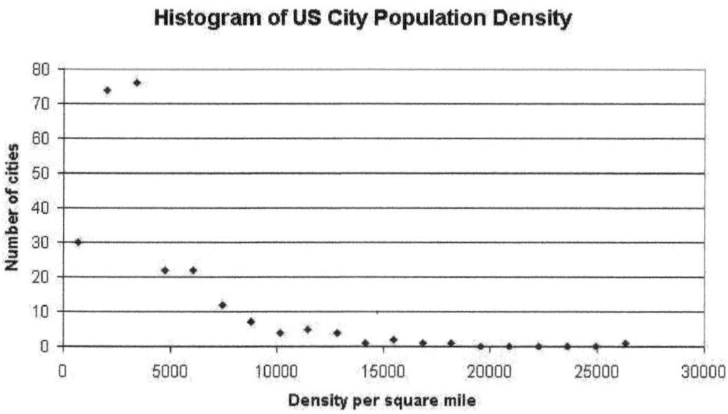

The mean density is 4128.21 inhabitants per square mile (1593.91 /km^2).

The median is 3160.85 inhabitants per square mile (1220.41 /km^2).

Population	Number of municipal governments[8]
1,000,000+	9
300,000–1,000,000	51
200,000–299,999	46
100,000–199,999	169
Total	**275**

U.S. Census Bureau statistical areas

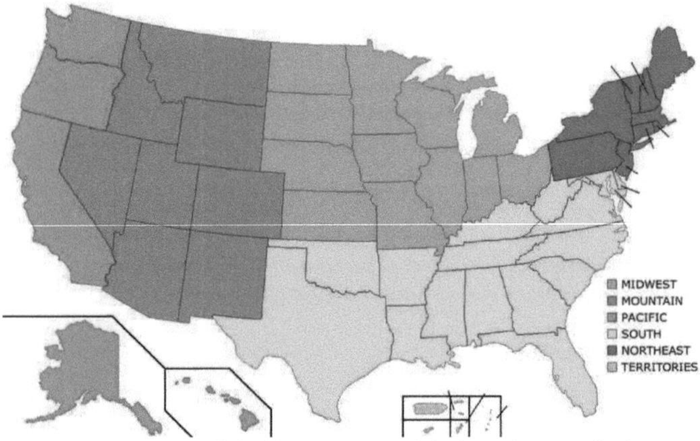

United States census statistical areas by state, district, or territory

AL

AK

AZ

AR

CA

CO

CT

DE

DC

FL

GA

HI

ID

IL

IN

IA

KS

KY

LA

ME

MD

MA

MI

MN

MS

MO

MT

NE

NV

NH

NJ

NM

NY

NC

ND

OH

OK

OR

PA

RI

SC

SD

TN

TX

UT

VT

VA

WA

WV

WI

WY

AS

MP

PR

GU

VI

References

[1] "American FactFinder" (http://factfinder2.census.gov). U.S. Census Bureau. .

[2] "U.S. Census Bureau Delivers Puerto Rico's 2010 Census Population Totals" (http://2010.census.gov/news/releases/operations/cb11-cn120.html). *2010 Census Population*. U.S. Census Bureau, Population Division. 2011-03-24. . Retrieved 2011-03-24.

[3] "Profile of General Population and Housing Characteristics: 2010" (http://factfinder2.census.gov/faces/tableservices/jsf/pages/productview.xhtml?pid=DEC_10_DP_DPDP1&prodType=table). United States Census Bureau, Population Division. . Retrieved 2011-07-28.

[4] "Table 1. Annual Estimates of the Resident Population for Counties of Indiana: April 1, 2000 to July 1, 2009 (CO-EST2009-01-18)" (http://www.census.gov/popest/counties/tables/CO-EST2009-01-18.csv) (CSV). U.S. Census Bureau, Population Division. March 2010. . Retrieved 2010-12-30.

[5] "Table 1. Annual Estimates of the Resident Population for Counties of Kentucky: April 1, 2000 to July 1, 2009 (CO-EST2009-01-21)" (http://www.census.gov/popest/counties/tables/CO-EST2009-01-21.csv) (CSV). U.S. Census Bureau, Population Division. March 2010. . Retrieved 2010-12-30.

[6] "Table 1. Annual Estimates of the Resident Population for Counties of Tennessee: April 1, 2000 to July 1, 2009 (CO-EST2009-01-47)" (http://www.census.gov/popest/counties/tables/CO-EST2009-01-47.csv) (CSV). U.S. Census Bureau, Population Division. March 2010. . Retrieved 2010-12-30.

[7] "Table 1. Annual Estimates of the Resident Population for Counties of Georgia: April 1, 2000 to July 1, 2009 (CO-EST2009-01-13)" (http://www.census.gov/popest/counties/tables/CO-EST2009-01-13.csv) (CSV). U.S. Census Bureau, Population Division. March 2010. . Retrieved 2010-12-30.

[8] "Census of Governments" (http://www.census.gov/prod/2003pubs/gc021x1.pdf). *Government Organization*. U.S. Census Bureau. December 2002. . Retrieved 30 December 2010.

External links

- "Biggest US Cities By Population" (http://www.biggestuscities.com/). *biggestuscities.com*. 11 December 2010.
- "Population Estimates: Incorporated Places and Minor Civil Divisions, by State (2000 to 2009)" (http://www.census.gov/popest/cities/SUB-EST2009-4.html). U.S. Census Bureau, Population Division. 2010.
- "U.S. Census Bureau Gazetteer (2010)" (http://www.census.gov/geo/www/gazetteer/files/Gaz_places_national.txt). U.S. Census Bureau.

British Columbia

British Columbia
***Colombie-Britannique* (French)**

Flag

Coat of arms

Motto: *Latin: Splendor sine occasu*
(English: *Splendour without diminishment*)

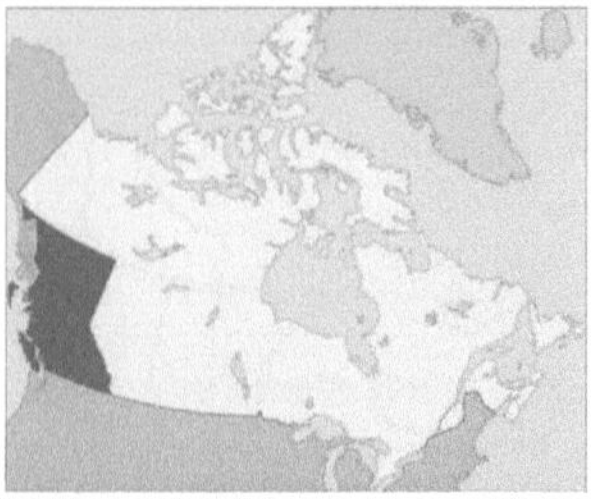

Capital	Victoria
Largest city	Vancouver
Largest metro	Metro Vancouver
Official languages	English (*de facto*, not *de jure*)
Demonym	British Columbian[1]
Government	
Type	Constitutional monarchy
Lieutenant-Governor	Steven Point
Premier	Christy Clark (BC Liberal)
Legislature	Legislative Assembly of British Columbia
Federal representation	In Canadian parliament
House seats	36 of 308 (11.7%)
Senate seats	6 of 105 (5.7%)
Confederation	20 July 1871 (7th)
Area	Ranked 5th
Total	944735 km^2 (sq mi)
Land	925186 km^2 (sq mi)
Water (%)	19549 km^2 (7548 sq mi) (2.1%)
Proportion of Canada	9.5% of 9,984,670 km^2
Population	Ranked 3rd

Total (2010)	4510858 (est.)[2]
Density (2010)	4.88 /km^2 (12.6 /sq mi)
GDP	Ranked 4th
Total (2009)	C$191.006 billion[3]
Per capita	C$41,689 (7th)
Abbreviations	
Postal	BC
ISO 3166-2	CA-BC
Time zone	UTC−8 & −7
Postal code prefix	V
Flower	Pacific Dogwood
Tree	Western Red Cedar
Bird	Steller's Jay
Website	www.gov.bc.ca [4]
Rankings include all provinces and territories	

British Columbia i/ˌbrɪtɪʃkəˈlʌmbiə/ (**B.C.** or **BC**) (French: *la Colombie-Britannique*, *C.-B.*) is the westernmost of Canada's provinces and is known for its natural beauty, as reflected in its Latin motto, *Splendor sine occasu* ("Splendour without Diminishment"). Its name was chosen by Queen Victoria in 1858. In 1871, it became the sixth province of Canada.

As well as being the westernmost province of Western Canada, British Columbia is also a component of the Pacific Northwest, along with the US states of Oregon and Washington. [5] [6]

The capital of British Columbia is Victoria, the 15th largest metropolitan region in Canada. The largest city is Vancouver, the third-largest metropolitan area in Canada, the largest in Western Canada, and the second largest in the Pacific Northwest. In 2009, British Columbia had an estimated population of 4,419,974 (about two and a half million of whom were in Greater Vancouver). The province is currently governed by the BC Liberal Party, led by Premier Christy Clark, who became leader as a result of the party election on February 26, 2011.

British Columbia's economy is largely resource-based. It is the endpoint of transcontinental highways and railways and the site of major Pacific ports, which enable international trade. Less than five percent of its land is arable; the province is agriculturally rich (particularly in the Fraser and Okanagan Valleys) because of its mild weather. Its climate encourages outdoor recreation and tourism, though its economic mainstay has long been resource extraction, principally logging and mining. While the coast of BC and certain valleys in the south-central part of the province have mild weather, the majority of BC's land mass experiences a cold winter temperate to subarctic climate similar to the rest of Canada.

Etymology

The province's name was chosen by Queen Victoria when the Colony of British Columbia, i.e. "the Mainland", became a British colony in 1858.[7] It refers to the Columbia District, the British name for the territory drained by the Columbia River, in southeastern British Columbia, which was the namesake of the pre-Oregon Treaty Columbia Department of the Hudson's Bay Company. Queen Victoria chose *British* Columbia to distinguish what was the British sector of the Columbia District from that of the United States ("American Columbia" or "Southern Columbia"), which became the Oregon Territory in 1848 as a result of the treaty.

Geography

British Columbia is bordered by the Pacific Ocean on the west, by the U.S. state of Alaska on the northwest as well as parts of the west, on the north by the Yukon and the Northwest Territories, on the east by the province of Alberta, and on the south by the U.S. states of Washington, Idaho, and Montana. The current southern border of British Columbia was established by the 1846 Oregon Treaty, although its history is tied with lands as far south as California. British Columbia's land area is 944735 square kilometres (sq mi). British Columbia's rugged coastline stretches for more than 27000 kilometres (17000 mi), and includes deep, mountainous fjords and about six thousand islands, most of which are uninhabited.

British Columbia's geography is epitomized by the variety and intensity of its physical relief, which has defined patterns of settlement and industry since colonization.

British Columbia's capital is Victoria, located at the southeastern tip of Vancouver Island. The province's most populous city is Vancouver, which is not on Vancouver Island but rather is located in the southwest corner of the mainland (an area often called the Lower Mainland). Other major cities include Surrey, Burnaby, Coquitlam, Richmond, Delta, and New Westminster in the Lower Mainland; Abbotsford, Pitt Meadows, Maple Ridge and Langley in the Fraser Valley; Nanaimo on Vancouver Island; and Kelowna and Kamloops in the Interior. Prince George is the largest city in the northern part of the province, while a village northwest of it, Vanderhoof, is near the geographic centre of the province.[8]

The Coast Mountains and the Inside Passage's many inlets provide some of British Columbia's renowned and spectacular scenery, which forms the backdrop and context for a growing outdoor adventure and ecotourism industry. Seventy-five percent of the province is mountainous (more than 1000 metres (3300 ft) above sea level); 60% is forested; and only about 5% is arable.

The Okanagan area is one of several wine-growing regions in Canada and also produces ciders; other wine regions in British Columbia include the Cowichan Valley on Vancouver Island, and the Fraser Valley. The cities of Kamloops and Penticton, and rural towns of Oliver, and Osoyoos have some of the warmest and longest summer climates in Canada, although their temperature ranges are exceeded by the warmer Fraser Canyon towns of Lillooet and Lytton, where shade temperatures on summer afternoons often surpass 40 °C (104 °F) but with very low humidity.

Much of the western part of Vancouver Island and the rest of the coast is covered by temperate rainforest. The province's mainland away from the coastal regions is not as moderated by the Pacific Ocean and ranges from desert and semi-arid plateau to the range and canyon districts of the Central and Southern Interior to boreal forest and sub-arctic prairie in the Northern Interior.

A few Southern Interior valleys have short cold winters with infrequent heavy snow, while those in the Cariboo, the southern part of the Central Interior, are colder because of their altitude and latitude, but without the intensity or duration experienced at similar latitudes elsewhere in Canada. The northern two-thirds of the province is largely unpopulated and undeveloped, and is mostly mountainous except east of the Rockies, where the Peace River District, in the northeast of the province contains BC's portion of the Canadian Prairies.

Climate

As a result of Kuroshio Current (also known as the Japan Current), which crosses the North Pacific Ocean, coastal British Columbia has a mild, rainy oceanic climate. Due to the blocking presence of successive mountain ranges, the Interior of the province has a semi-arid climate with certain locations receiving less than 250mm (10") in annual precipitation. The annual mean temperature in the most populated areas of the province are above 10 °C (50 °F), the mildest anywhere in Canada.

Strait of Georgia, near Vancouver.

Winters can be severe in the Interior and the North. For example, the average overnight low in Prince George (roughly located in the middle of the province) in January is −14 °C (7 °F). The coldest temperature in British Columbia was recorded in Smith River, where it dropped to −58.9 °C (−74 °F), one of the coldest readings recorded anywhere in North America. Southern Interior valleys have shorter winters with brief bouts of cold. Heavy snowfall occurs in the Coast, Columbia and Rocky Mountains providing healthy bases for skiers.

On the Coast, rainfall, sometimes relentless heavy rain, dominates in winter because of consistent barrages of cyclonic low-pressure systems from the North Pacific, but on occasion (and not every winter) heavy snowfalls and below freezing temperatures arrive when modified arctic air reaches coastal areas for typically short periods. On the opposite extreme, summers in the Southern Interior valleys are hot, for example in Osoyoos the July Maximum averages 32 °C (90 °F), hot weather sometimes moves towards the Coast or to the far North. Temperatures have gone over 40 °C (104 °F) in the past, with the record high being held in Lytton, when the temperature rose to 44.4 °C (111.9 °F) on July 16, 1941.[9]

The extended summer dryness often creates conditions that spark forest fires, from dry-lightning or man-made causes. Coastal areas are generally milder and dry during summer, under the influence of stable anti-cyclonic high pressure much of the time. Many areas of the province are often covered by a blanket of heavy cloud and low fog during winter, despite sunny summers. Annual sunshine hours vary from 2200 near Cranbrook and Victoria to less than 1300 sun hours per year in Prince Rupert, located on the North Coast, just south of the Alaska Panhandle.

Parks and protected areas

There are 14 designations of parks and protected areas in the province that reflects the different administration and creation of these areas in a modern context. There are 141 ecological Reserves, 35 provincial marine parks, 7 Provincial Heritage Sites, 6 National Historic Sites of Canada, 4 National Parks and 3 National Park Reserves. 12.5% (114000 km^2 (44000 sq mi)) of British Columbia is currently considered protected under one of the 14 different designations that includes over 800 distinct areas.

Mount Robson, Canadian Rockies, B.C.

British Columbia contains seven of Canada's national parks:

- Glacier National Park
- Gulf Islands National Park Reserve
- Gwaii Haanas National Park Reserve and Haida Heritage Site
- Kootenay National Park
- Mount Revelstoke National Park
- Pacific Rim National Park Reserve
- Yoho National Park

British Columbia also contains a large network of provincial parks, run by BC Parks of the Ministry of Environment. British Columbia's provincial parks system is the second largest parks system in Canada (the largest is Canada's National Parks system).

Another tier of parks in British Columbia are regional parks, which are maintained and run by regional districts.

In addition to these areas, over 47000 km^2 (18000 sq mi) of arable land are protected by the Agricultural Land Reserve.

Fauna

Much of the province is wild or semi-wild, so that populations of many mammalian species that have become rare in much of the United States still flourish in British Columbia. Watching animals of various sorts, including a very wide range of birds, has also long been popular. Bears (grizzly, black, and the Kermode bear or spirit bear—found only in British Columbia) live here, as do deer, elk, moose, caribou, big-horn sheep, mountain goats, marmots, beavers, muskrat, coyotes, wolves, mustelids (such as wolverines, badgers and fishers), Cougar, eagles, ospreys, herons, Canada geese, swans, loons, hawks, owls, ravens, Harlequin Ducks, and many other sorts of ducks. Smaller birds (robins, jays, grosbeaks, chickadees, and so on) also abound.

Yoho National Park

Healthy populations of many sorts of fish are found in the waters (including salmonids such as several species of salmon, trout, char, and so on.). Besides salmon and trout, sport-fishers in B.C. also catch halibut, steelhead, bass, and sturgeon. On the coastlines, Harbor Seals and river otters are common. Cetacean species native to the coast include the Orca, Gray Whale, Harbour Porpoise, Dall's Porpoise, Pacific White-sided Dolphin and Minke Whale.

British Columbian introduced species include: common dandelion, ring-necked pheasant, Pacific oyster, brown trout, black slug, European Starling, cowbird, knapweed, bullfrog, purple loosestrife, Scotch broom, European earwig, tent caterpillar, sowbug, gray squirrel, Asian longhorn beetle, English ivy, Fallow Deer, thistle, gorse, Norway rat, crested mynah, and Asian or European gypsy moth.

Some endangered species in British Columbia are: Vancouver Island marmot, Spotted Owl, American White Pelican, and badgers.

Cheakamus Lake at dawn, in Garibaldi Provincial Park, B.C.

Type of organism	Red-listed species in BC	Total number of species in BC
Freshwater fish	24	80
Amphibians	5	19
Reptiles	6	16
Birds	34	465
Terrestrial mammals	Requires new data	Requires new data
Marine mammals	3	29
Plants	257	2333
Butterflies	12	187

Dragonflies	9	35000

As of 2001[10]

Ecozones

Environment Canada subdivides British Columbia into a system of 6 ecozones:

- Pacific Marine
- Pacific Maritime
- Boreal Cordillera
- Montane Cordillera
- Taiga Plains
- Boreal Plains Ecozones.

History

Fur trade and colonial era

The discovery of stone tools on the Beatton River near Fort St. John date human habitation in British Columbia to at least 11,500 years ago. The Indigenous peoples of the Pacific Northwest Coast spread throughout the region, achieving a high population density; at the time of European contact, nearly half the aboriginal people in present-day Canada lived in the region. During the 1770s, smallpox killed at least 30% of the Pacific Northwest First Nations.[11] This epidemic was the first and the most devastating of a number that were to follow, other than the Great Smallpox Epidemic of 1862 which killed off 50% of the native population in that year.[12]

The arrival of Europeans began around the mid-18th century, as fur traders entered the area to harvest sea otters. While it is thought that Sir Francis Drake may have explored the British Columbian coast in 1579, it was Juan Francisco de la Bodega y Quadra who completed the first documented voyage, which took place in 1775. In doing so, Quadra reasserted the Spanish claim for the Pacific coast, first made by Vasco Núñez de Balboa in 1513. The explorations of James Cook in 1778 and George Vancouver in 1792-93 established British jurisdiction over the coastal area north and west of the Columbia River. In 1793, Sir Alexander Mackenzie was the first European to journey across North America overland to the Pacific Ocean, inscribing a stone marking his accomplishment on the shoreline of Dean Channel near Bella Coola. His expedition theoretically established British sovereignty inland, and a succession of other fur company explorers charted the maze of rivers and mountain ranges between the Canadian Prairies and the Pacific. Mackenzie and these other explorers—notably John Finlay, Simon Fraser, Samuel Black, and David Thompson—were primarily concerned with extending the fur trade, rather than political considerations. In 1794, by the third of a series of agreements known as the Nootka Conventions, Spain conceded its claims of exclusivity in the Pacific. This opened the way for formal claims and colonization by other powers, including Britain, but because of the Napoleonic Wars there was little British action on its claims in the region until later.

Kwakwaka'wakw house pole, second half of the 19th century

The establishment of trading posts under the auspices of the North West Company and the Hudson's Bay Company (HBC), effectively established a permanent British presence in the region. The Columbia District, broadly defined as being south of 54°40 north latitude, (the southern limit of Russian America) and north of Mexican Controlled California west of the Rocky Mountains was, by the Anglo-American Convention of 1818, under the "joint occupancy and use" of citizens of the United States and subjects of Britain (which is to say, the fur companies). This co-occupancy was ended with the Oregon Treaty of 1846.

The major supply route was the York Factory Express between Hudson Bay and Fort Vancouver. Some of the early outposts grew into settlements, communities, and cities. Among the places in British Columbia that began as fur trading posts are Fort St. John (established 1794); Hudson's Hope (1805); Fort Nelson (1805); Fort St. James (1806); Prince George (1807); Kamloops (1812); Fort Langley (1827); Fort Victoria (1843); Yale (1848); and Nanaimo (1853). Fur company posts that became cities in what is now the United States include Vancouver, Washington (Fort Vancouver), formerly the "capital" of Hudson's Bay operations in the Columbia District, Colville, Washington and Walla Walla, Washington (old Fort Nez Percés).

With the amalgamation of the two fur trading companies in 1821, the region now comprising British Columbia existed in three fur trading departments. The bulk of the central and northern interior was organized into the New Caledonia district, administered from Fort St. James. The interior south of the Thompson River watershed and north of the Columbia was organized into the Columbia District, administered from Fort Vancouver on the lower Columbia River. The northeast corner of the province east of the Rockies, known as the Peace River Block, was attached to the much larger Athabasca District, headquartered in Fort Chipewyan, in present day Alberta.

Until 1849, these districts were a wholly unorganized area of British North America under the de facto jurisdiction of HBC administrators. Unlike Rupert's Land to the north and east, however, the territory was not a concession to the company. Rather, it was simply granted a monopoly to trade with the First Nations inhabitants. All that was changed with the westward extension of American exploration and the concomitant overlapping claims of territorial sovereignty, especially in the southern Columbia basin (within present day Washington and Oregon). In 1846, the

Oregon Treaty divided the territory along the 49th parallel to the Georgia Strait, with the area south of this boundary (excluding Vancouver Island and the Gulf Islands) transferred to sole American sovereignty. The Colony of Vancouver Island was created in 1849, with Victoria designated as the capital. New Caledonia, as the whole of the mainland rather than just its north-central Interior came to be called, continued to be an unorganized territory of British North America, "administered" by individual HBC trading post managers.

With the Fraser Canyon Gold Rush in 1858, an influx of Americans into New Caledonia prompted the colonial office to formally designate the mainland as the Colony of British Columbia, with New Westminster as its capital. A series of gold rushes in various parts of the province followed, the largest being the Cariboo Gold Rush in 1862, forcing the colonial administration into deeper debt as it struggled to meet the extensive infrastructure needs of far-flung boom communities like Barkerville and Lillooet, which sprang up overnight. The Vancouver Island colony was facing financial crises of its own, and pressure to merge the two eventually succeeded in 1866.

Rapid growth and development

Lord Strathcona drives the Last Spike of the Canadian Pacific Railway, at Craigellachie, 7 November 1885. Completion of the transcontinental railroad was a condition of entry into Confederation.

The Confederation League, including such figures as Amor De Cosmos, John Robson, and Robert Beaven, led the chorus pressing for the colony to join Canada, which had been created out of three British North American colonies in 1867 (the Province of Canada, Nova Scotia and New Brunswick). Several factors motivated this agitation, including the fear of annexation to the United States, the overwhelming debt created by rapid population growth, the need for government-funded services to support this population, and the economic depression caused by the end of the gold rush.

With the agreement by the Canadian government to extend the Canadian Pacific Railway to British Columbia and to assume the colony's debt, British Columbia became the sixth province to join Confederation on 20 July 1871. The borders of the province were not completely settled until 1903, however, when the province's territory shrank somewhat after the Alaska boundary dispute settled the vague boundary of the Alaska Panhandle.

Population in British Columbia continued to expand as the province's mining, forestry, agriculture, and fishing sectors were developed. Mining activity was particularly notable throughout the Mainland, particularly in the Boundary Country, in the Slocan, in the West Kootenay around Trail, the East Kootenay (the southeast corner of the province), the Fraser Canyon, the Cariboo, the Omineca and the Cassiar, so much so a common epithet for the Mainland, even after provincehood, was "the Gold Colony". Agriculture attracted settlers to the fertile Fraser Valley, and cattle ranchers and later fruit growers came to the drier grasslands of the Thompson River area, the Cariboo, the Chilcotin, and the Okanagan. Forestry drew workers to the lush temperate rainforests of the coast, which was also the locus of a growing fishery.

The completion of the railway in 1885 was a huge boost to the province's economy, facilitating the transportation of the region's considerable resources to the east. The milltown of Granville, known as Gastown, near the mouth of the Burrard Inlet was selected as the terminus of the railway, prompting the incorporation of the City as Vancouver in 1886. The completion of the Port of Vancouver spurred rapid growth, and in less than fifty years the city surpassed Winnipeg, Manitoba, as the largest in Western Canada. The early decades of the province were ones in which issues of land use—specifically, its settlement and development—were paramount. This included expropriation from First Nations people of their land, control over its resources, as well as the ability to trade in some resources (such as the fishery).

Establishing a labour force to develop the province was problematic from the start, and British Columbia was the locus of immigration from Europe, China, and Japan. The influx of a non-Caucasian population stimulated resentment from the dominant ethnic groups, resulting in agitation (much of it successful) to restrict the ability of Asian people to immigrate to British Columbia through the imposition of a head tax. This resentment culminated in mob attacks against Chinese and Japanese immigrants in Vancouver in 1887 and 1907. By 1923, almost all Chinese immigration had been blocked except for merchants, professionals, students and investors.

Meanwhile, the province continued to grow. In 1914, the last spike of a second transcontinental rail line, the Grand Trunk Pacific, linking north-central British Columbia from the Yellowhead Pass through Prince George to Prince Rupert was driven at Fort Fraser. This opened up the North Coast and the Bulkley Valley region to new economic opportunities. What had previously been an almost exclusively fur trade and subsistence economy soon became a locus for forestry, farming, and mining.

1920s through 1940s

When the men returned from World War I, they discovered the recently enfranchised women of the province had helped vote in the prohibition of liquor in an effort to end the social problems associated with the hard-core drinking that Vancouver and the rest of the province was famous for until the war. Because of pressure from veterans, prohibition was quickly relaxed so that the "soldier and the working man" could enjoy a drink, but widespread unemployment among veterans was hardened by many of the available jobs being taken by European immigrants and disgruntled veterans organized a range of "soldier parties" to represent their interests, variously named Soldier-Farmer, Soldier-Labour, and Farmer-Labour Parties. These formed the basis of the fractured labour-political spectrum that would generate a host of fringe leftist and rightist parties, including those who would eventually form the Co-operative Commonwealth and the early Social Credit splinter groups.

Internment camp for the Japanese Canadians during World War II

The advent of prohibition in the United States created new opportunities, and many found employment or at least profit in cross-border liquor smuggling. Much of Vancouver's prosperity and opulence in the 1920s results from this "pirate economy", although growth in forestry, fishing and mining continued. The end of U.S. prohibition, combined with the onset of the Great Depression, plunged the province into economic destitution. Compounding the already dire local economic situation, tens of thousands of men from colder parts of Canada swarmed into Vancouver, creating huge hobo jungles around False Creek and the Burrard Inlet rail yards, including the old Canadian Pacific Railway mainline right-of-way through the heart of the city's downtown (at Hastings and Carrall). Increasingly desperate times led to intense political organizing efforts, an occupation of the main Post Office at Granville and Hastings which was violently put down by the police and an effective imposition of martial law on the docks for almost three years. A Vancouver contingent for the On-to-Ottawa Trek was organized and seized a train, which was loaded with thousands of men bound for the capital but was met by a Gatling gun straddling the tracks at Mission; the men were arrested and sent to work camps for the duration of the Depression.[13]

There were some signs of economic life beginning to return to normal towards the end of the 1930s, but it was the onset of World War II which transformed the national economy and ended the hard times of the Depression. Because of the war effort, women entered the workforce as never before.

British Columbia has long taken advantage of its location on the Pacific Ocean to have close relations with East Asia. However, this has often caused friction between cultures which have caused occasional displays of animosity toward Asian immigrants. This was most manifest during the Second World War when many people of Japanese descent were relocated or interned in the Interior of the province.

Coalition and the post-War boom

The BC Regiment (DCOR), marching in New Westminster, 1940.

During World War II the mainstream BC Liberal and BC Conservative Parties of British Columbia united in a formal coalition government under new Liberal leader John Hart, who replaced Duff Pattullo when the latter failed to win a majority in the 1941 election. While the Liberals won the most number of seats, they actually received fewer votes than the socialist Co-operative Commonwealth Federation (CCF). Pattullo was unwilling to form a coalition with the rival Conservatives led by Royal Maitland and was replaced by Hart who formed a coalition cabinet made up of five Liberal and three Conservative ministers.[14] The CCF was invited to join the coalition but refused.[14]

The pretext for continuing the coalition after the end of World War II was to prevent the CCF, which had won a surprise victory in Saskatchewan in 1944, from ever coming to power in British Columbia. The CCF's popular vote was high enough in the 1945 election that they were likely to have won three-way contests and could have formed government. However, the coalition prevented that by uniting the anti-socialist vote.[14] In the post-war environment the government initiated a series of infrastructure projects, notably the completion of Highway 97 north of Prince George to the Peace River Block, a section called the John Hart Highway and also public hospital insurance.

In 1947 the reins of the Coalition were taken over by Byron Ingemar Johnson. The Conservatives had wanted their new leader Herbert Anscomb to be premier, but the Liberals in the Coalition refused. Johnson led the coalition to the highest percentage of the popular vote in British Columbia history (61%) in the 1949 election. This victory was attributable to the popularity of his government's spending programmes, despite rising criticism of corruption and abuse of power. During his tenure, major infrastructure continued to expand, and the agreement with Alcan to build the Kemano-Kitimat hydro and aluminum complex was put in place. Johnson achieved popularity for flood relief efforts during the 1948 flooding of the Fraser Valley, which was a major blow to that region and to the province's economy.

Increasing tension between the Liberal and Conservative coalition partners led the Liberal Party executive to vote to instruct Johnson to terminate the arrangement. Johnson ended the coalition and dropped his Conservative cabinet ministers, including Deputy Premier and Finance Minister Herbert Anscomb, precipitating the general election of 1952.[14] A referendum on electoral reform prior to this election had instigated an elimination ballot (similar to a preferential ballot), where voters could select second and third choices. The intent of the ballot, as campaigned for by Liberals and Conservatives, was that their supporters would list the rival party in lieu of the CCF, but this plan backfired when a large group of voters from all major parties, including the CCF, voted for the fringe British Columbia Social Credit Party (Socreds), who wound up with the largest number of seats in the House (19), only one seat ahead of the CCF, despite the CCF having 34.3% of the vote to Social Credit's 30.18%.

The Social Credit Party, led by rebel former Conservative MLA W. A. C. Bennett, formed a minority government backed by the Liberals and Conservatives (with 6 and 4 seats respectively). Bennett began a series of fiscal reforms, preaching a new variety of populism as well as waxing eloquent on progress and development, laying the ground for a second election in 1953 in which the new Bennett regime secured a majority of seats, with 38% of the vote. Secure with that majority, Bennett returned the province to the first-past-the-post system thereafter, which is still in use today.

Growth of government in the economy

Premier W. A. C. Bennett and his wife accompany Princess Margaret in Victoria, August 1958. Bennett's Socreds governed the province for an unprecedented twenty years

With the election of the Social Credit Party, British Columbia embarked a phase of rapid economic development. Bennett and his party governed the province for the next twenty years, during which time the government initiated an ambitious programme of infrastructure development, fuelled by a sustained economic boom in the forestry, mining, and energy sectors.

During these two decades, the government nationalized British Columbia Electric and the British Columbia Power Company, as well as smaller electric companies, renaming the entity BC Hydro. By the end of the 1960s, several major dams had been begun or completed in—among others—the Peace, Columbia, and Nechako River watersheds. Major transmission deals were concluded, most notably the Columbia River Treaty between Canada and the United States. The province's economy was also boosted by unprecedented growth in the forest sector, as well as oil and gas development in the province's northeast.

The 1950s and 1960s were also marked by development in the province's transportation infrastructure. In 1960, the government established BC Ferries as a crown corporation, in order to provide a marine extension of the provincial highway system. That system was improved and expanded through the construction of new highways and bridges, and paving of existing highways and provincial roads.

Vancouver and Victoria become cultural centres as poets, authors, artists, musicians, as well as dancers, actors, and *haute cuisine* chefs flocked to the beautiful scenery and warmer temperatures. Similarly, these cities have either attracted or given rise to their own noteworthy academics, commentators, and creative thinkers. Tourism also began to play an important role in the economy. The rise of Japan and other Pacific economies was a great boost to British Columbia's economy.

Politically and socially, the 1960s brought a period of significant social ferment. The divide between the political left and right, which had prevailed in the province since the Depression and the rise of the labour movement, sharpened as so-called free enterprise parties coalesced into the defacto coalition represented by Social Credit—in opposition to the social democratic New Democratic Party, the successor to the Co-operative Commonwealth Federation. As the province's economy blossomed, so did labour-management tensions. Tensions emerged, also, from the counterculture movement of the late 1960s, of which Vancouver and Nanaimo were centres. The conflict between hippies and Vancouver mayor Tom Campbell was particularly legendary, culminating in the so-called Gastown Riots of 1971. By the end of the decade, with social tensions and dissatisfaction with the status quo rising, the Bennett government's achievements could not stave off its growing unpopularity.

1970s and 1980s

On August 27, 1969, the Social Credit Party was re-elected in a general election for what would be Bennett's final term in power. At the start of the 1970s, the economy was quite strong because of rising coal prices and an increase in annual allowable cuts in the forestry sector. However, BC Hydro reported its first loss, which was the beginning of the end for Bennett and the Social Credit Party.[15]

The Socreds were forced from power in the August 1972 election, paving the way for a provincial New Democratic Party (NDP) government under Dave Barrett. Under Barrett, the large provincial surplus soon became a deficit, although changes to the accounting system makes it likely that some of the deficit was carried over from the previous Social Credit regime and its "two sets of books", as WAC Bennett had once referred to his system of fiscal management. The brief three year ("Thousand Days") period of NDP governance brought several lasting changes to the province, most notably the creation of the Agricultural Land Reserve, intended to protect farmland from redevelopment, and the Insurance Corporation of British Columbia, a crown corporation charged with a monopoly on providing single-payer basic automobile insurance.

Perceptions that the government had instituted reforms either too swiftly or that were too far-reaching, coupled with growing labour disruptions led to the ouster of the NDP in the 1975 general election. Social Credit, under W.A.C. Bennett's son, Bill Bennett, was returned to office. Under the younger Bennett's government, 85% of the province's land base was transferred from Government Reserve to management by the Ministry of Forests, reporting of deputy ministers was centralized to the Premier's Office, and NDP-instigated social programs were rolled back, with then-Human Resources Minister infamously demonstrating a golden shovel to highlight his welfare policy, although the new-era Socreds also reinforced and backed certain others instigated by the NDP—notably the creation of the Resort Municipality of Whistler, whose special status including Sunday drinking, then an anomaly in BC.

Also during the "MiniWac" regime (a reference to his father's acronym-cum-nickname, WAC) certain money-losing Crown-owned assets were "privatized" in a mass giveaway of shares in the British Columbia Resources Investment Corporation, "BCRIC", with the "Brick shares" soon becoming near-worthless. Towards the end of his tenure in power, Bennett oversaw the completion of several megaprojects meant to stimulate the economy and win votes - unlike most right-wing parties, British Columbia's Social Credit actively practiced government stimulation of the economy. Most notable of these was the winning of a world's fair for Vancouver, which came in the form of Expo 86, to which was tied the construction of the Coquihalla Highway and Vancouver's SkyTrain system. The Coquihalla Highway project became the subject of a scandal after revelations that the premier's brother bought large tracts of land needed for the project before it was announced to the public, and also because of graft investigations of the huge cost overruns on the project. Both investigations were derailed in the media by a still further scandal, the Doman Scandal, in which the Premier and millionaire backer Herb Doman were investigated for insider-trading and securities fraud. Nonetheless, the Socreds were re-elected in 1979 under Bennett, who led the party until 1986.

The Coquihalla Highway. The road was one of the legacies of the Expo 86 World's fair. The creation of the toll highway sparked controversy.

As the province entered a sustained recession, Bennett's popularity and media image were in decline. On April 1, 1983, Premier Bennett overstayed his constitutional limits of power by exceeding the legal tenure of a government, and the Lieutenant-Governor, Henry Pybus Bell-Irving, was forced to call Bennett to Government House to resolve the impasse, and an election was called for April 30, while in the meantime government cheques were covered by special emergency warrants as the Executive Council no longer had signing authority because of the constitutional crisis. Campaigning on a platform of moderation, and backed by the support and computer-organization tactics of the Big Blue Machine from Ontario and other consultants who were electoral lobbyists for the American Republican Party, Bennett won an unexpected majority.

After several weeks of silence in the aftermath, a sitting of the House was finally called and in the speech from the Throne the Socreds instituted a programme of fiscal cutbacks dubbed "restraint", which had been a buzzword for moderation during the campaign. The programme included cuts to "motherhood" issues of the left, including the human rights branch, the offices of the Ombudsman and Rentalsman, women's programs, environmental and cultural programs, while still supplying mass capital infusions to corporate British Columbia. This sparked a backlash, with tens of thousands of people in the streets the next day after the budget speech, and through the course of a summer repeated large demonstrations of up to 100,000 people.

This became known as the 1983 Solidarity Crisis, from the name of the Solidarity Coalition, a huge grassroots opposition movement mobilized, consisting of organized labour and community groups, with the British Columbia Federation of Labour forming a separate organization of unions, Operation Solidarity, under the direction of Jack Munro, then-President of the International Woodworkers of America (IWA), the most powerful of the province's resource unions. Tens of thousands participated in protests and many felt that a general strike would be the inevitable result unless the government backed down from its policies they had claimed were only about restraint and not about recrimination against the NDP and the left. Just as a strike at Pacific Press ended, which had crippled the political management of the public agenda by the publishers of the province's major papers, the movement collapsed after an apparent deal was struck by union leader and IWA president, Jack Munro and Premier Bennett.[16]

A tense winter of blockades at various job sites around the province ensued, as among the new laws were those enabling non-union labour to work on large projects and other sensitive labour issues, with companies from Alberta and other provinces brought in to compete with union-scale British Columbia companies. Despite the tension, Bennett's last few years in power were relatively peaceful as economic and political momentum grew on the megaprojects associated with Expo, and Bennett was to end his career by hosting Prince Charles and Lady Diana on their visit to open Expo 86. His retirement being announced, a Social Credit convention was scheduled for the Whistler Resort, which came down to a three-way shooting match between Bud Smith, the Premier's right-hand man but an unelected official, Social Credit party grande dame Grace McCarthy, and the charismatic but eccentric Bill Vander Zalm.

Bill Vander Zalm became the new Socred leader when Smith threw his support to him rather than see McCarthy win, and led the party to victory in the election later that year. Vander Zalm was later involved in a conflict of interest scandal following the sale of Fantasy Gardens, a Christian and Dutch culture theme park built by the Premier, to Tan Yu, a Filipino Chinese gambling kingpin. There were also concerns over Yu's application to the government for a bank licence, and lurid stories from flamboyant realtor Faye Leung of a party in the "Howard Hughes Suite" on the top two floors of the Bayshore Inn, where Tan Yu had been staying, with reports of a bag of money in a brown paper bag passed from Yu to Vander Zalm during the goings-on. These scandals forced Vander Zalm's resignation, and Rita Johnston became premier of the province. Johnston presided over the end of Social Credit power, calling an election which led to the reducing of the party's caucus to only two seats, and the revival of the long-defunct British Columbia Liberal Party as Opposition to the victorious NDP under former Vancouver mayor Mike Harcourt.

In 1988, David Lam was appointed by the Queen of Canada as British Columbia's twenty-fifth Lieutenant-Governor, and was the Province's first Lieutenant-Governor of Chinese origin.

1990s to present

Johnston lost the 1991 general election to the NDP, under the leadership of Mike Harcourt, a former mayor of Vancouver. The NDP's unprecedented creation of new parkland and protected areas was popular, and helped boost the province's growing tourism sector. However, the economy continued to struggle against the backdrop of a weak resource economy. Housing starts and an expanded service sector saw growth overall through the decade, despite political turmoil. Harcourt ended up resigning over "Bingogate"—a political scandal involving the funnelling of charity bingo receipts into party coffers in certain ridings. Harcourt was not implicated, but he resigned nonetheless in respect of constitutional conventions calling for leaders under suspicion to step aside. Glen Clark, a former president of the BC Federation of Labour, was chosen the new leader of the NDP, which won a second term in 1996. More scandals dogged the party, most notably the Fast Ferry Scandal involving the province trying to develop the shipbuilding industry in British Columbia. An allegation (never substantiated) that the Premier had received a favour in return for granting a gaming licence led to Clark's resignation as Premier. He was succeeded on an interim basis by Dan Miller who was in turn followed by Ujjal Dosanjh following a leadership convention.

In the 2001 general election Gordon Campbell's BC Liberals defeated the NDP party, gaining 77 out of 79 seats total seats in the provincial legislature. Campbell instituted various reforms and removed some of the NDP's policies including scrapping the "fast ferries" project, lowering income taxes, and the controversial sale of BC Rail to CN Rail. Campbell was also the subject of criticism after he was arrested for driving under the influence during a vacation in Hawaii. However, Campbell still managed to lead his party to victory in the 2005 general election, against a substantially strengthened NDP opposition. Campbell won a third term in the British Columbia general election, 2009, marking the first time in 23 years that a premier has been elected to a third term.

The province successfully won a bid to host the 2010 Winter Olympics in Vancouver and Whistler, with Olympic organizers winning a referendum held in the city of Vancouver.

British Columbia has also been significantly affected by demographic changes within Canada and around the world. Vancouver (and to a lesser extent some other parts of British Columbia) was a major destination for many of the immigrants from Hong Kong who left the former UK colony (either temporarily or permanently) in the years immediately prior to its handover to the People's Republic of China. British Columbia has also been a significant destination for internal Canadian migrants. This has been the case throughout recent decades, because of its image of natural beauty, mild climate and relaxed lifestyle, but is particularly true during periods of economic growth. As a result, British Columbia has moved from approximately 10% of Canada's population in 1971 to approximately 13% in 2006. Trends of urbanization mean that the Greater Vancouver area now includes 51% of the Province's population, followed in second place by Greater Victoria with 8%. These two metropolitan regions have traditionally dominated the demographics of BC.

Demographics

Welcome sign at the province's boundary

Year	Population	Five Year % change	Ten Year % change	Rank Among Provinces
1851	55,000	n/a	n/a	6
1861	51,524	n/a	−6.3	6
1871	36,247	n/a	−35.3	7
1881	49,459	n/a	36.4	8
1891	98,173	n/a	98.5	8
1901	178,657	n/a	82.0	6
1911	392,480	n/a	119.7	6
1921	524,582	n/a	33.7	6
1931	694,263	n/a	32.3	6
1941	817,861	n/a	17.8	3
1951	1,165,210	n/a	42.5	3
1956	1,398,464	20.0	n/a	3

1961	1,629,082	16.5	39.8	3
1966	1,873,674	15.0	34.0	3
1971	2,184,620	16.6	34.1	3
1976	2,466,610	12.9	31.6	3
1981	2,744,467	11.3	25.6	3
1986	2,883,370	5.1	16.9	3
1991	3,282,061	13.8	19.6	3
1996	3,724,500	13.5	29.2	3
2001	3,907,738	4.9	19.1	3
2006	4,113,487	5.3	10.4	3

[17] [18]

Religious groups in BC (1991 & 2001) & Canada (2001)

	1991 BC %	**2001 BC %**	**2001 Canada %**	**BC 2001 number**	
Total population	100%	100%	100%	3,868,875	
No religious affiliation	30.0%	35.1%	16.5%	1,388,300	includes Agnostic, Atheist, Humanist, and No religion, and other responses, such as Darwinism, and so on.
Protestant	41.9%	31.4%	29.2%	1,213,295	
Catholic	18.3%	17.2%	43.6%	675,320	includes Roman Catholic, Eastern Catholic .
Christian Orthodox	0.7%	0.9%	1.6%	35,655	
Christian n. i. e.	2.7%	5.2%	2.6%	200,345	Includes mostly answers of 'Christian', not otherwise stated
Sikh	2.3%	3.5%	0.9%	135,310	
Buddhist	1.1%	2.2%	1%	85,540	
Muslim	0.8%	1.5%	2%	56,220	
Hindu	0.6%	0.8%	1%	31,500	
Jewish	0.5%	0.5%	1.1%	21,230	
Eastern religions		0.3%	0.1%	9,970	includes Baha'i, Eckankar, Jains, Shinto, Taoist, Zoroastrian and Eastern religions, not identified elsewhere
Other religions		0.4%	0.2%	16,205	includes Aboriginal spirituality, Pagan, Wicca, Unity – New Thought – Pantheist, Scientology, Rastafarian, New Age, Gnostic,and so on.

[19] [20]

The largest denominations by number of adherents according to the 2001 census were none (atheist, agnostic, and so on.) with 1,388,300 (35.9%); protestant with 1,213,295 (31.4%); the Roman Catholic Church with 675,320 (17%); the United Church of Canada with 361,840 (9%); and the Anglican Church of Canada with 298,375 (8%).[21]

Ethnic groups and languages

The following statistics represent both single (for example, "German") and multiple (for example, "Chinese-Canadian") responses to the 2006 Census, and thus do not add up to 100%. All items are self-identified.[22]

Ethnic Origin	Population	Percent
English	1,207,245	29.6%
Scottish	828,145	20.3%
Canadian	720,200	17.7%
Irish	618,120	15.2%
German	561,570	13.8%
Chinese	432,435	10.6%
French	361,215	8.9%
East Indian	232,370	5.7%
Ukrainian	197,265	4.8%
Dutch (Netherlands)	196,420	4.8%
North American Indian	193,060	4.7%
Italian	143,155	3.5%
Norwegian	129,420	3.2%
Polish	128,360	3.2%
Russian	114,105	2.8%
Welsh	104,275	2.6%
Swedish	104,025	2.6%
Filipino	94,255	2.3%
British	74,145	1.8%
American (USA)	66,765	1.6%

Ethnic Origin	Population	Percent
Métis	62,570	1.5%
Danish	56,125	1.4%
Spanish	52,640	1.3%
Korean	51,860	1.3%
Hungarian	49,870	1.2%
Austrian	46,620	1.1%
Japanese	41,585	1.0%
Portuguese	34,660	0.9%
Vietnamese	30,835	0.8%
Jewish	30,830	0.8%
Finnish	29,875	0.7%
Iranian	29,265	0.7%
Swiss	28,240	0.7%

Romanian	25,670	0.6%
Icelandic	22,110	0.5%
Greek	21,770	0.5%
Czech	21,150	0.5%
Croatian	18,815	0.5%
Punjabi	18,525	0.5%
Belgian	17,510	0.4%

Though just 1.8% refer to their origins as being "British", a majority 57.3% of the population of British Columbia claim their ethnic origin as being from one of the British nations (England, Scotland or Wales). Another 15.2% refer to their ethnicity as "Irish", though not distinguishing between Northern Ireland and the Republic of Ireland.

Of the provinces, British Columbia had the highest proportion of visible minorities, representing 24.8% of its population.[23] Asians are by far the largest visible minority demographic, with many of the Lower Mainland's large cities having sizable Chinese, South Asian, Japanese, Filipino, and Korean communities.

Also present in large numbers relative to other regions of Canada (except Toronto), and ever since the province was first settled (unlike Toronto), are many European ethnicities of the first and second generation, notably Germans, Scandinavians, Yugoslavs and Italians. Third-generation Europeans are generally of mixed lineage, and traditionally intermarried with other ethnic groups more than in any other Canadian province. First-generation Britons remain a strong component of local society despite limitations on immigration from Britain since the ending of special status for British subjects in the 1960s.

Of the 4,113,847 population counted by the 2006 census, 4,074,385 people completed the section about language. Of these 4,022,045 gave singular responses to the question regarding mother tongue. The languages most commonly reported were the following:

Language	Number of native speakers	Percentage of singular responses
English	2,875,770	71.5%
Chinese languages	342,920	8.5%
Punjabi	158,750	4.0%
German	86,690	2.2%
French	54,745	1.4%
Tagalog (Filipino)	50,425	1.3%
Korean	46,500	1.2%
Spanish	34,075	0.9%
Persian	28,150	0.7%
Italian	27,020	0.7%
Dutch	26,355	0.7%
Vietnamese	24,560	0.7%
Hindi	23,240	0.6%
Japanese	20,040	0.5%
Russian	19,320	0.5%
Polish	17,565	0.4%
Portuguese	14,385	0.4%

Ukrainian	12,285	0.3%
Hungarian	10,670	0.3%
Croatian	8,505	0.2%

Language	Number of native speakers	Percentage of singular responses
Arabic	8,440	0.2%
Urdu	7,025	0.2%
Danish	6,720	0.2%
Greek	6,620	0.2%
Gujarati	6,565	0.2%
Romanian	6,335	0.2%
Serbian	6,180	0.2%
Czech	6,000	0.1%
Finnish	4,770	0.1%
Athabaskan languages	3,500	0.1%
Slovak	3,490	0.1%
Norwegian	3,275	0.1%
Tamil	3,200	0.1%
Salish languages	3,190	0.1%
Ilocano	3,100	0.1%
Malay	3,100	0.1%
Bisayan languages	3,035	0.1%
Swedish	2,875	0.1%
Turkish	2,255	0.1%
Tsimshianic languages	2,125	0.1%

Numerous other languages were also counted, but only languages with more than 2,000 native speakers are shown. (Figures shown are for the number of single language responses and the percentage of total single-language responses)[24]

Cities

Vancouver Skyline

Half of all British Columbians live in the Metro Vancouver area, which includes Vancouver, Surrey, New Westminster, West Vancouver, North Vancouver (city), North Vancouver (district municipality), Burnaby, Coquitlam, Port Coquitlam, Maple Ridge, Langley (city), Langley (district municipality), Delta, Pitt Meadows, White Rock, Richmond, Port Moody, Anmore, Belcarra, Lions Bay and Bowen Island, with adjacent unincorporated areas represented in the regional district as the electoral area known as Greater Vancouver Electoral Area A. Seventeen Indian reserves are located in the metropolitan area but are outside the jurisdiction of the regional district and not represented in its government. Also in the metropolitan area but not represented in the regional district are the University Endowment Lands.

The second largest concentration of British Columbia population is located at the southern tip of Vancouver Island, which is made up of the 13 municipalities of Greater Victoria, Victoria, Saanich, Esquimalt, Oak Bay, View Royal, Highlands, Colwood, Langford, Central Saanich/Saanichton, North Saanich, Sidney, Metchosin, Sooke, which are part of the Capital Regional District. The metropolitan area also includes several Indian reserves (the governments of which are not part of the regional district). Almost half of the Vancouver Island population is located in Greater Victoria.

Ten Largest Metropolitan Areas in BC by Population[25] [26]

Community (includes metro areas)	2006	1996
Vancouver	2,215,200	1,831,665
Victoria	330,088	304,287
Kelowna	162,276	136,349
Abbotsford	159,020	136,480
Kamloops	92,882	85,407
Nanaimo	92,361	82,691
Prince George	83,225	87,731
Fort St. John	81,000	68,000
Chilliwack	80,892	66,254
Vernon	55,418	49,701
Courtenay	49,214	46,297

Ten Largest Municipalities in BC by Population

Municipality	2006	1996
Vancouver	578,041	514,008
Surrey (Metro Vancouver)	394,976	304,477
Burnaby (Metro Vancouver)	202,799	179,209
Richmond (Metro Vancouver)	174,461	148,867
Abbotsford	123,864	104,403
Coquitlam (Metro Vancouver)	114,565	101,820
Saanich (Greater Victoria)	108,265	101,388
Kelowna	106,707	89,422
Delta (Metro Vancouver)	96,723	95,411
Langley Township (Metro Vancouver)	93,726	80,179

Other municipalities include: Campbell River, Chilliwack, Colwood, Courtenay, Cranbrook, Dawson Creek, Fernie, Fort St. John, Kimberley, Langford, Maple Ridge, Mission, Parksville, North Cowichan, Penticton, Port Alberni, Prince George, Prince Rupert, Quesnel, Terrace, Vernon, Victoria (provincial capital), Williams Lake

Economy

A view of the Vancouver's downtown core. Vancouver is the business capital of British Columbia.

British Columbia has a resource dominated economy, centred on the forestry industry but also with increasing importance in mining. Employment in the resource sector has fallen steadily, and new jobs are mostly in the construction and retail/service sectors. With its film industry known as Hollywood North, the Vancouver region is the third-largest feature film production location in North America, after Los Angeles and New York City.[27]

The economic history of British Columbia is replete with tales of dramatic upswings and downswings, and this boom and bust pattern has influenced the politics, culture and business climate of the province. Economic activity related to mining in particular has widely fluctuated with changes in commodity prices over time, with documented costs to community health.[28]

British Columbia's GDP is the fourth largest in Canada at C$197.93 billion in 2008. GDP per capita stands at C$45,150.[29] British Columbia's total debt will rise 16% to C$47.8 billion in the 2010-11 fiscal year, or 24.3% of GDP.[30]

Transportation

Transportation played a huge role in British Columbia history. The Rocky Mountains and the ranges west of them constituted a significant obstacle to overland travel until the completion of the transcontinental railway in 1885. The Peace River Canyon through the Rocky Mountains was the route that the earliest explorers and fur traders used. Fur trade routes were only marginally used for access to British Columbia through the mountains. Travel from the rest of Canada before 1885 meant the difficulty of overland travel via the United States, around Cape Horn or overseas from Asia. Nearly all travel and freight to and from the region occurred via the Pacific Ocean, primarily through the ports of Victoria and New Westminster.

Until the 1930s, rail was the only means of overland travel to and from the rest of Canada; travellers using motor vehicles needed to journey through the United States. With the construction of the Inter-Provincial Highway in 1932 (now known as the Crowsnest Pass Highway), and later the Trans-Canada Highway, road transportation evolved into the preferred mode of overland travel to and from the rest of the country.

Roads and highways

Alex Fraser Bridge on Highway 91 in Richmond/Delta.

Because of its size and rugged, varying topography, British Columbia requires thousands of kilometres of provincial highways to connect its communities. British Columbia's roads systems were notoriously poorly maintained and dangerous until a concentrated programme of improvement was initiated in the 1950s and 1960s. There are now freeways in the Lower Mainland and Central Interior of the province, and much of the rest of the province is accessible by well-maintained two lane arterial highways with additional passing lanes in mountainous areas. The building and maintenance of provincial highways is the responsibility of the provincial government.

There are only five major routes to the rest of Canada. From south to north they are: BC Highway 3 through the Crowsnest Pass, the Vermilion Pass and the Kicking Horse Pass, the latter being used by the Trans-Canada Highway entering Alberta through Banff National Park, the Yellowhead Highway through Jasper National Park, and Highway

2 through Dawson Creek. There are also several highway crossings to the adjoining American states of Washington, Idaho, and Montana. The longest highway is Highway 97, running 2081 km (1293 mi) from the British Columbia-Washington border at Osoyoos north to Watson Lake, Yukon and which includes the British Columbia portion of the Alaska Highway.

Public transit

Prior to 1979, surface public transit was administered by BC Hydro, the provincially owned electricity utility. Subsequently, the province established BC Transit to oversee and operate all municipal transportation systems. In 1998, the Greater Vancouver Transportation Authority (TransLink) (now South Coast British Columbia Transportation Authority), a separate authority for routes within the Greater Vancouver Regional District was established.

Public transit in British Columbia consists mainly of diesel buses, although Vancouver is also serviced by a fleet of trolleybuses. Several experimental buses are being tested such as hybrid buses that has both gasoline and electric engines. TransLink operates SkyTrain, an automated metro system serving the cities of Vancouver, Burnaby, New Westminster, North Surrey and Richmond. In 2009, the Canada Line SkyTrain was completed, linking Vancouver International Airport to downtown Vancouver bringing the total to three operating metro lines.

Presently, a new line to Coquitlam and Port Moody (the Evergreen Line) being designed, expansion of the Expo Line is being developed, and there is planning for an extension of the Millennium Line through Vancouver City to the University of British Columbia. Also of note, Turnstiles are being added to all existing (and future) lines of the system. In the past, SkyTrain used a proof of payment honour system.

Rail

CPR train traversing the Stoney Creek Bridge.

Rail development expanded greatly in the decades after the completion of the Canadian Pacific Railway in 1885 and was the chief mode of long-distance surface transportation until the expansion and improvement of the provincial highways system began in the 1950s. Two major routes through the Yellowhead Pass competed with the Canadian Pacific Railway—the Grand Trunk Pacific, terminating at Prince Rupert, and the Canadian National Railway, terminating at Vancouver.

The Pacific Great Eastern line supplemented this service, providing a north-south route between Interior resource communities and the coast. The Pacific Great Eastern (later known as British Columbia Railway and now owned by Canadian National Railway) connects Fort St James, Fort Nelson, and Tumbler Ridge with North Vancouver. The E&N Railway, rebranded as Southern Railway of Vancouver Island, serves the commercial and passenger train markets of Vancouver Island by owning the physical rail lines.

Amtrak runs twice daily trains between Vancouver and Seattle.

Water

BC Ferries was established as a provincial crown corporation in 1960 to provide passenger and vehicle ferry service between Vancouver Island and the Lower Mainland as a cheaper and more reliable alternative to the service operated by the Canadian Pacific Railway. It now operates 25 routes among the islands of British Columbia, as well as between the islands and the mainland. Ferry service to Washington is offered by the Washington State Ferries (between Sidney and Anacortes) and Black Ball Transport (between Victoria and Port Angeles, Washington). Ferry service over inland lakes and rivers is provided by the provincial government.

Commercial ocean transport is of vital importance. Major ports are located at Vancouver, Roberts Bank (near Tsawwassen), Prince Rupert, and Victoria. Of these, the Port of Vancouver is the most important, being the largest in Canada and the most diversified in North America. Vancouver, Victoria, and Prince Rupert are also major ports of call for cruise ships. In 2007, a large maritime container port was opened in Prince Rupert with an inland sorting port located in Prince George.

Air

There are over 200 airports located throughout British Columbia, the major ones being the Vancouver International Airport, the Victoria International Airport, the Kelowna International Airport, and the Abbotsford International Airport, the first three of which each served over 1,000,000 passengers in 2005. Vancouver International Airport is the 2nd busiest airport in the country and the second biggest International Gateway on the west coast (after Los Angeles) with an estimated 17.9 million travellers passing through in 2008.

Government and politics

The Lieutenant-Governor of British Columbia, Steven Point, is the Queen of Canada's representative in the Province of British Columbia. During the absence of the Lieutenant-Governor, the Governor General in Council may appoint an administrator to execute the duties of the office. In practice, this is usually the Chief Justice of British Columbia.[31]

The British Columbia Parliament Buildings in Victoria

British Columbia has an 85-member elected Legislative Assembly, elected by the plurality voting system, though in recent years there has been significant debate about switching to a single transferable vote system.

The province is governed by the Liberal Party under Premier Christy Clark. The previous election saw *then* Premier Gordon Campbell win a third straight majority government in May 2009, taking 49 seats to the opposition New Democratic Party's 35. Campbell had previously led the largest landslide election in British Columbia history in 2001, with 77 of 79 seats, but the legislature has been more evenly divided between Liberals and NDP following the 2005 (46 of 79) and 2009 (49 of 85) provincial elections. The New Democratic Party (NDP) and its predecessor the CCF have been the main opposition force to business-oriented parties since the 1930s and have ruled majority governments 1972-1975 and 1991-2001. The Green Party of British Columbia plays a larger role in the politics of British Columbia than Green Parties do in most other jurisdictions in Canada. However, after a breakthrough election in 2001 (12.39%), the party's vote share has declined (2005 – 9.17%, 2009 – 8.09%).

The British Columbia Liberal Party is not related to the federal Liberal Party and does not share the same ideology. Instead, the BC Liberal party is a rather diverse coalition, made up of the remnants of the Social Credit Party, many federal Liberals, federal Conservatives, and those who would otherwise support right-of-centre or free enterprise parties. Historically, there have commonly been third parties present in the legislature (including the Liberals themselves from 1952 to 1975), but there are presently none.

Prior to the rise of the Liberal Party, British Columbia's main political party was the British Columbia Social Credit Party which ruled British Columbia for 20 continuous years. While sharing some ideology with the current Liberal government, they were more right-wing although undertook nationalization of various important monopolies, notably BC Hydro and BC Ferries. In an April 2008 poll by polling firm Ipsos-Reid, the BC Liberals were shown as having the support of 49% of voters, compared to 32% for the NDP.[32]

The meeting chamber of the provincial legislative assembly

British Columbia is known for having politically active labour unions who have traditionally supported the NDP or its predecessor, the CCF.

British Columbia's political history is typified by scandal and a cast of colourful characters, beginning with various colonial-era land scandals and abuses of power by early officials (such as those that led to McGowan's War in 1858–59). Notable scandals in Social Credit years included the Robert Bonner Affair, the Fantasy Gardens scandal which forced Premier Bill Vander Zalm to resign and ended the Social Credit era, the Bingogate scandal which brought down NDP Premier Mike Harcourt, the alleged scandal named Casinogate which drove NDP Premier Glen Clark to resign. A variety of scandals have plagued the current Liberal government, but with little apparent effect on the electorate, including the Premier's arrest for drunk driving in Maui and the resignation of various cabinet ministers because of conflict-of-interest allegations. A Christmas Eve raid on the Parliament Buildings in Victoria, including the Premier's Office, has resulted in charges only for ministerial aides, although key cabinet members from the time have since resigned. The case, currently in preliminary hearings in the courts and relating to the sale of BC Rail to an American company, may not reach trial because of the mass of evidence and various procedural problems.

Recreation

Given its varied mountainous terrain and its coasts, lakes, rivers, and forests, British Columbia has long been enjoyed for pursuits like hiking and camping, rock climbing and mountaineering, hunting and fishing.

Ice sailing in Whistler

Water sports, both motorized and non-motorized, are enjoyed in many places. Sea kayaking opportunities abound on the British Columbia coast with its fjords. Whitewater rafting and kayaking are popular on many inland rivers. Sailing and sailboarding are widely enjoyed.

In winter, cross-country and telemark skiing are much enjoyed, and in recent decades high-quality downhill skiing has been developed in the Coast Mountain range and the Rockies, as well as in the southern areas of the Shuswap Highlands and the Columbia Mountains. Snowboarding has mushroomed in popularity since the early 1990s. The 2010 Winter Olympics downhill events were held in Whistler Blackcomb area of the province, while the indoor events were conducted in the Vancouver area.

In Vancouver and Victoria (as well as some other cities), opportunities for joggers and bicyclists have been developed. Cross-country bike touring has been popular since the ten-speed bike became available many years ago. Since the advent of the more robust mountain bike, trails in more rugged and wild places have been developed for them. Some of the province's retired rail beds have been converted and maintained for hiking, biking, and cross-country skiing. Longboarding is also a popular activity because of the hilly geography of the region.

Horseback riding is enjoyed by many British Columbians. Opportunities for trail riding, often into especially scenic areas, have been established for tourists in numerous areas of the province.

British Columbia also has strong participation levels in many other sports, including golf, tennis, soccer, hockey, Canadian football, rugby union, lacrosse, baseball, softball, basketball, curling and figure skating. British Columbia has produced many outstanding athletes, especially in aquatic and winter sports.

A crop of Cannabis Sativa, or "BC Bud".

Consistent with both increased tourism and increased participation in diverse recreations by British Columbians has been the proliferation of lodges, chalets, bed and breakfasts, motels, hotels, fishing camps, and park-camping facilities in recent decades.

In certain areas, there are businesses, non-profit societies, or municipal governments dedicated to promoting ecotourism in their region. A number of British Columbia farmers offer visitors to combine tourism with farm work, for example, through the WWOOF Canada program.[33]

Recreational cannabis

A 2004 study (published 2006) by the University of Victoria Centre for Addictions Research of BC and Simon Fraser University Applied Research on Mental Health and Addictions indicated cannabis use is more widespread among British Columbians than other Canadians.[34] However, a United Nations report published in July 2007 actually placed Quebec as the highest consumption province, citing 15.8% of Quebecers having used marijuana in a single year, versus 14.1% of Canadians nationally,[35] and resulted in Canada being placed first in the industrialized world in marijuana use. With the actual growing of marijuana, British Columbia is responsible for 40% of all cannabis produced in Canada.[36]

See also

- List of designated places in British Columbia
- List of people from British Columbia
- List of regions of Canada: British Columbia

References

[1] According to the *Oxford Guide to Canadian English Usage* (ISBN 0-19-541619-8; p. 335), *BCer(s)* is an informal demonym that is sometimes used for residents of BC.

[2] "Canada's population estimates: Table 2 Quarterly demographic estimates" (http://www.statcan.gc.ca/daily-quotidien/100628/t100628a2-eng.htm). Statcan.gc.ca. 2010-06-28. . Retrieved 2010-06-30.

[3] "Statistics Canada Gross domestic product, expenditure-based, by province and territory" (http://www40.statcan.ca/l01/cst01/econ15.htm). .

[4] http://www.gov.bc.ca/

[5] "Pacific Northwest Washington Oregon British Columbia" (http://www.celebratebig.com/pacific-northwest/index.htm). Celebratebig.com. . Retrieved 2012-01-24.

[6] "Northwest Hiking Trails - Washington, Oregon, Idaho" (http://www.cooltrails.com/). Cooltrails.com. . Retrieved 2012-01-24.

[7] Ged Martin, "The Naming of British Columbia," *Albion: A Quarterly Journal Concerned with British Studies,* Vol. 10, No. 3 (Autumn, 1978), pp. 257–263 in JSTOR (http://www.jstor.org/stable/4048132)

[8] "Vanderhoof" (http://www.hellobc.com/en-CA/RegionsCities/Vanderhoof.htm). Tourism BC. . Retrieved 2007-04-26.

[9] "Daily Data Report for July 1941" (http://climate.weatheroffice.gc.ca/climateData/dailydata_e.html?timeframe=2&Prov=CA&StationID=966&Year=1941&Month=7&Day=26). Climate.weatheroffice.gc.ca. 2010-09-21. . Retrieved 2011-02-22.

[10] BC Ministry of Sustainable Resource Management, Conservation Data Centre

[11] Lange, Greg (2003-01-23). "Smallpox epidemic ravages Native Americans on the northwest coast of North America in the 1770s." (http://www.historylink.org/essays/output.cfm?file_id=5100). Historylink.org. . Retrieved 2011-02-22.

[12] "Plagues and Peoples on the Northwest Coast" (http://www2.h-net.msu.edu/reviews/showrev.php?id=4547). .h-net.msu.edu. . Retrieved 2011-02-22.

[13] *Carved From Wood; Mission BC 1861-1992*, Andreas Schroeder, publ. The Mission Foundation (1992) ISBN 1550561316m, ISBN 978-1550561319, ASIN: B000B8XZNA

[14] Hans J. Michelmann, David E. Smith, Cristine De Clercy *Continuity And Change in Canadian Politics: Essays in Honour of David E. Smith* (http://books.google.com/books?id=d7iLYjWHvS8C&pg=PA184&lpg=PA184&dq=herbert+anscomb+conservative#PPA184,M1), University of Toronto Press (2006), page 184

[15] Elections BC (1998). "Electoral History of British Columbia 1871–1986" (http://web.archive.org/web/20061020072258/http://www.llbc.leg.bc.ca/public/PubDocs/bcdocs/37907/electoral_history.pdf). Archived from the original (http://www.llbc.leg.bc.ca/public/PubDocs/bcdocs/37907/electoral_history.pdf) on 2006-10-20. . Retrieved 2007-04-26.

[16] Palmer, Bryan (1987). *Solidarity: The Rise and Fall of an Opposition in British Columbia*. Vancouver: New Star Books. ISBN.

[17] "Statistics Canada — Population" (http://www40.statcan.ca/l01/cst01/demo62k.htm). .

[18] Canada's population (http://www.statcan.ca/Daily/English/060927/d060927a.htm). Statistics Canada. Retrieved 2006-09-28.

[19] Statistics Canada (http://www12.statcan.ca/english/census01/Products/Analytic/companion/rel/tables/provs/bcmajor.cfm).

[20] Statistics Canada (http://www12.statcan.ca/english/census01/products/standard/prprofile/prprofile.cfm?G=59).

[21] "Religions in Canada" (http://www12.statcan.ca/english/census01/products/highlight/Religion/Page.cfm?Lang=E&Geo=PR&View=1a&Code=59&Table=1&StartRec=1&Sort=2&B1=59&B2=All). 2.statcan.ca. . Retrieved 2011-02-22.

[22] "2006 Canadian Census" (http://www12.statcan.ca/english/census06/data/topics/RetrieveProductTable.cfm?ALEVEL=3&APATH=3&CATNO=&DETAIL=0&DIM=&DS=99&FL=0&FREE=0&GAL=0&GC=99&GK=NA&GRP=1&IPS=&METH=0&ORDER=1&PID=92333&PTYPE=88971&RL=0&S=1&SUB=0&ShowAll=No&StartRow=1&Temporal=2006&Theme=80&VID=0&VNAMEE=&VNAMEF=&GID=838062). .

[23] "Statistics Canada. "Canada's Ethnocultural Mosaic, 2006 Census: Provinces and territories"" (http://www12.statcan.ca/english/census06/analysis/ethnicorigin/british.cfm).

[24] *Detailed Mother Tongue (148), Single and Multiple Language Responses (3) and Sex (3) for the Population of Canada, Provinces, Territories, Census Metropolitan Areas and Census Agglomerations, 2006 Census – 20% Sample Data* (http://www12.statcan.ca/english/census06/data/topics/RetrieveProductTable.cfm?ALEVEL=3&APATH=3&CATNO=&DETAIL=0&DIM=&DS=99&FL=0&FREE=0&GAL=0&GC=99&GK=NA&GRP=1&IPS=&METH=0&ORDER=1&PID=89186&PTYPE=88971&RL=0&S=1&ShowAll=No&StartRow=1&SUB=701&Temporal=2006&Theme=70&VID=0&VNAMEE=&VNAMEF=&GID=838062). 2007. .

[25] Statistics Canada (2002). Statistics Canada "Population and Dwelling Counts, for Canada, Provinces and Territories, Census Metropolitan Areas and Census Agglomerations" (http://www12.statcan.ca/english/census01/products/standard/popdwell/Table-CMA-P.cfm?T=1&SR=1&PR=59&S=3&O=D). Statistics Canada. Retrieved 2007-04-26.

[26] Indian reserve populations are not included in these figures

[27] Vancouver Economic Development (2005). "Film and TV" (http://web.archive.org/web/20070101014505/http://vancouvereconomic.com/key_sectors/film_tv.htm). Archived from Film and Development the original (http://www.vancouvereconomic.com/key_sectors/film_tv.htm) on 2007-01-01. . Retrieved 2007-04-26.

[28] University of British Columbia (September 2006). "Hard on Health of Mining Communities" (http://www.publicaffairs.ubc.ca/ubcreports/2005/05oct06/mining.html). . Retrieved 2007-04-26.

[29] "Gross domestic product, expenditure-based, by province and territory" (http://www40.statcan.gc.ca/l01/cst01/econ15-eng.htm). 0.statcan.gc.ca. 2009-11-10. . Retrieved 2010-06-30.

[30] " British Columbia sets budget, eyes Olympic costs (http://www.reuters.com/article/idUSN0218838520100302)". Reuters. March 2, 2010.

[31] Executive Power in the Provinces (http://lois.justice.gc.ca/en/const/c1867_e.html#provincial) under the Constitutional Act, 1867.

[32] S. 23 of the Constitution Act (British Columbia) requires elections to be held on the second Tuesday in May every fourth year after May 2005. Constitution Act (http://www.bclaws.ca/Recon/document/freeside/-- C --/Constitution Act RSBC 1996 c.66/00_96066_01.xml#section23). Retrieved 2009-04-06.

[33] "WWOOF — Willing Workers on Organic Farms, Canada" (http://wwoof.ca/). .

[34] "Cannabis Use Highest in BC" (http://communications.uvic.ca/releases/release.php?display=release&id=758). University of Victoria, Victoria, British Columbia. 2006-10-04. . Retrieved 2012-01-17.

[35] Gazette, The (2007-07-14). "Quebec gone to pot" (http://www.canada.com/montrealgazette/news/story.html?id=89dd1b3f-eaf2-4945-99d9-de420f19dbd5&k=64745). Canada.com. . Retrieved 2012-01-17.

[36] Vancouver, The (2007-07-10). "Canada leads 'rich' world in using marijuana: UN" (http://www.canada.com/vancouversun/news/story.html?id=21fd2469-4720-417e-b4ff-9e889f5588e8). Canada.com. . Retrieved 2012-01-17.

- . BC Geographical Names. http://apps.gov.bc.ca/pub/bcgnws/names/.html.

Further reading

- Akrigg, G. P. V; Akrigg, Helen B (1997). *British Columbia place names* (http://books.google.ca/books?id=9IP4V-Hyt-4C&lpg=PP1&dq=British Columbia&pg=PP1#v=onepage&q&f=true). Univ. of British Columbia Press. ISBN 0774806362.
- Barman, Jean (2007). *The West beyond the West: a history of British Columbia* (http://books.google.ca/books?id=JbYe6fCOSTAC&lpg=PA3&dq=British Columbia&pg=PP1#v=onepage&q&f=true). Univ. of Toronto Press. ISBN 0802093094.
- Dawson, Michael (2005). *Selling British Columbia: Tourism and Consumer Culture, 1890-1970* (http://books.google.ca/books?id=LMzgxL7WhzAC&lpg=PP1&dq=British Columbia&pg=PP1#v=onepage&q&f=true). Univ. of British Columbia Press. ISBN 9780774810555.
- Foster, Leslie T (2007). *People, politics, and child welfare in British Columbia* (http://books.google.ca/books?id=c598qbT6gBMC&lpg=PA251&dq=British Columbia&pg=PP1#v=onepage&q&f=true). Univ. of British Columbia Press. ISBN 9780774813723.
- Harris, R Cole (1997). *The resettlement of British Columbia : essays on colonialism and geographical change* (http://books.google.ca/books?id=TzvH3MCYTm4C&lpg=PP1&dq=British Columbia&pg=PP1#v=onepage&q&f=true). Univ. of British Columbia Press. ISBN 0774805889.
- McKee, Christopher (2000). *Treaty talks in British Columbia: negotiating a mutually beneficial future* (http://books.google.ca/books?id=51FVXiNA7okC&lpg=PA120&dq=British Columbia&pg=PP1#v=onepage&q&f=true). Univ. of British Columbia Press. ISBN 0774808241.
- McGillivray, Brett (2000). *Geography of British Columbia: people and landscapes in transition* (http://books.google.ca/books?id=MWL_6b7dUrkC&lpg=PP1&dq=British Columbia&pg=PP1#v=onepage&q&f=true). Univ. of British Columbia Press. ISBN 0774807849.
- Muckle, Robert James (1998). *The First Nations of British Columbia: an anthropological survey* (http://books.google.ca/books?id=8zBQ951iVncC&lpg=PP1&dq=British Columbia&pg=PP1#v=onepage&q&f=true). Univ. of British Columbia Press. ISBN 077480663X.
- Thirkell, Fred; Bob Scullion (2002). *British Columbia 100 years ago: portraits of a province* (http://books.google.ca/books?id=rPbtpsZF5aEC&lpg=PP1&dq=British Columbia&pg=PP1#v=onepage&q&f=true). Heritage House Pub. ISBN 1894384490.
- Ver Berkmoes, Ryan; John Lee (2007). *Guide to "British Columbia"* (http://books.google.ca/books?id=yfFTSXftIOcC&lpg=PP1&dq=British Columbia&pg=PP1#v=onepage&q&f=true). Lonely Planet. ISBN 9781741045840.

External links

- Tourism British Columbia official website (http://www.hellobc.com/)
- British Columbia (http://www.dmoz.org/Regional/North_America/Canada/British_Columbia/) at the Open Directory Project
- Provincial Archives including online photo database (http://www.bcarchives.gov.bc.ca/)
- Vancouver Public Library; Historical Photographs of BC & the Yukon (http://www.vpl.vancouver.bc.ca/find/cat/C393/)
- B.C. Multicultural Photographs from the Vancouver Public Library – searchable photo database (http://www.multiculturalcanada.ca/vpl)
- BC Govt online map archive (http://maps.gov.bc.ca/)

Portland metropolitan area

Portland-Vancouver-Hillsboro, OR-WA Metropolitan Statistical Area

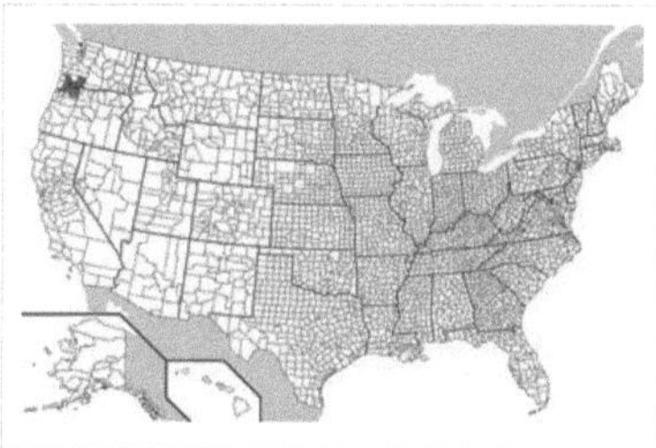

Common name: Portland metropolitan area

Largest city	Portland, Oregon (583,776)
Other cities	Vancouver (161,791) Gresham (105,594) Hillsboro (91,611)
Population	Ranked 23rd in the U.S.
- Total	2,226,009
- Density	333/sq. mi. 129/km²
Area	6,684 sq. mi. 17,310 km²
State(s)	Oregon Washington
Elevation	
- Highest point	N/A feet (N/A m)
- Lowest point	N/A feet (N/A m)

The **Portland-Vancouver-Hillsboro, OR-WA Metropolitan Statistical Area** (MSA), also known as the **Portland metropolitan area** or **Greater Portland**, is an urban area in the U.S. states of Oregon and Washington centered around the city of Portland, Oregon. The U.S. Census Bureau's definition for the MSA is currently all of the following counties: Clackamas County, Oregon; Columbia County, Oregon; Multnomah County, Oregon; Washington County, Oregon; Yamhill County, Oregon; Clark County, Washington; and Skamania County, Washington.[1] The Washington counties are separated from Oregon by the Columbia River.

The Oregon portion of the metropolitan area is the state's largest urban center. Portions of this are under the jurisdiction of Metro,[2] a directly elected regional government which, among other things, is responsible for land use planning in the region.

Cities and other communities

Major cities in the region in addition to Portland include Beaverton, Gresham, Hillsboro in Oregon, and Vancouver in Washington. The area also includes the smaller cities of Cornelius, Damascus, Fairview, Forest Grove, Gladstone, King City, Lake Oswego, Milwaukie, Oregon City, Sherwood, Tigard, Troutdale, Tualatin, West Linn, Wilsonville, Wood Village in Oregon, as well as Battle Ground, Camas, and Washougal in Washington.

It includes the unincorporated suburban communities in Oregon of Aloha, Beavercreek, Cedar Mill, Clackamas, Dunthorpe, Garden Home, Raleigh Hills, and West Slope.

Categorized lists of cities, neighborhoods, and communities are also available:

- Category:Multnomah County, Oregon
- Category:Washington County, Oregon
- Category:Clackamas County, Oregon
- Category:Clark County, Washington
- Category:Neighborhoods in Portland, Oregon

Metropolitan statistical area

The Portland–Vancouver–Hillsboro Metropolitan Statistical Area (MSA), the 23rd largest in the United States,[3] [4] has a population of 2,226,009 (2010 Census). Of them, 1,789,580 live in Oregon (46.7% of the state's population) while the rest 436,429 live in Washington (6.7% of state's population). It consists of Multnomah, Washington, Clackamas, Columbia and Yamhill counties in Oregon, as well as Clark and Skamania counties in Washington. The area includes Portland and the neighboring cities of Beaverton, Gresham, Hillsboro, Milwaukie, Lake Oswego, Oregon City, Fairview, Wood Village, Troutdale, Tualatin, Tigard, West Linn, Vancouver, Camas and Washougal.

Before the recent (2003–2004) redefinition of metropolitan boundaries, the Portland Consolidated Metropolitan Area included the Salem Metropolitan Statistical Area.[5] If still considered a part of the metropolitan area the population would be 2,637,944 (2009 estimate).[6] Combined population close connected Portland, Salem and Longview metropolitan areas is 2,739,910 (2009 estimate).[6]

Transportation

Portland is where Interstate 84 ends at Interstate 5, both major highways in the Pacific Northwest. Other primary roads include Interstate 205, an eastern bypass of the urban core, U.S. Route 26, which heads west and southeast, U.S. Route 30, which goes to the northwest, and Oregon Route 217, which connects US 26 with I-5 in the south, travelling through Beaverton. Both US 26 and US 30 go to the Oregon Coast. Also, Washington State Route 14 travels along the northern banks of the Columbia River from downtown Vancouver east to Camas and Washougal.

Transit service on the Oregon side is generally provided by TriMet. In addition, Sandy Area Metro serves Sandy, South Clackamas Transportation District serves nearby Molalla, Canby Area Transit serves Canby and South Metro Area Regional Transit serves Wilsonville. Service in Clark County is provided by C-TRAN. In Columbia County, the *Columbia County Rider* provides transit service on weekdays connecting St. Helens with downtown Portland and connecting Scappoose and St. Helens with certain points in urban Washington County, including the PCC Rock Creek campus, Tanasbourne and the Willow Creek MAX light rail station.[7]

References

[1] "Update of Statistical Area Definitions and Guidance on Their Uses" (http://www.whitehouse.gov/omb/bulletins/fy2008/b08-01.pdf) (PDF). Office of Management and Budget. November 20, 2007. pp. 45. . Retrieved 2008-09-05.

[2] "Jurisdictional Boundaries" (http://www.oregonmetro.gov/index.cfm/go/by.web/id=2948). Metro. . Retrieved 2011-08-01.

[3] http://www.census.gov/popest/metro/tables/2008/CBSA-EST2008-01.xls

[4] "OMB Bulletin No. 10-02: Update of Statistical Area Definitions and Guidance on Their Uses" (http://www.whitehouse.gov/omb/assets/bulletins/b10-02.pdf). United States Office of Management and Budget. December 1, 2009. . Retrieved January 18, 2010.

[5] http://www.demographia.com/db-porugbmigr.pdf

[6] http://www.census.gov/popest/metro/tables/2009/CBSA-EST2009-01.csv

[7] "Schedules & Routes" (http://www.columbiacountyrider.com/Schedules_Routes.html). Columbia County Rider. . Retrieved 2009-10-10.

External links

- Metro government website (http://www.metro-region.org/)
- Portland MSA 2010 Census numbers from the Population Resource Center (http://www.pdx.edu/sites/www.pdx.edu.prc/files/media_assets/2010_PL94_county_MSA_updated.pdf)
- pdx.edu/media/p/r/PRC_2007_Population_Report2_rev.pdf (http://www.pdx.edu/media/p/r/PRC_2007_Population_Report2_rev.pdf) of key urban planning documents on the Portland Metropolitan area, at Portland State University

Seattle

Seattle

— City —

City of Seattle

Flag

Seal

Nickname(s): The Emerald City, Seatown, Rain City, Jet City, Gateway to Alaska, Gateway to The Pacific, Queen City, The Town

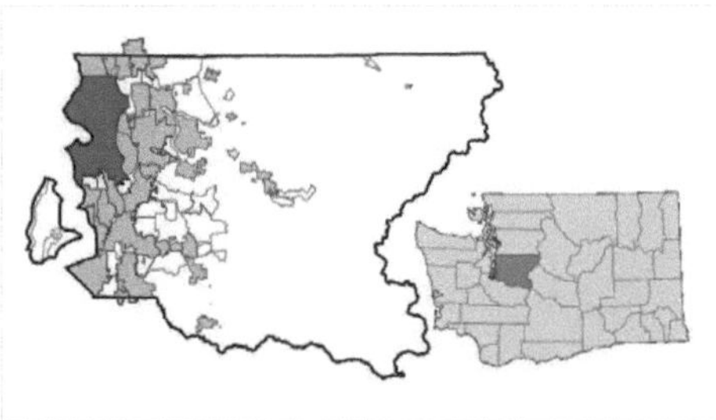

Location of Seattle in
King County and Washington

Seattle

Location in the United States

Coordinates: 47°36′35″N 122°19′59″W	
Country	USA
State	Washington
County	King
Incorporated	December 2, 1869
Government	
• **Mayor**	Michael McGinn
• **City Council**	**List of Councilors**
Area	
• **City**	142.5 sq mi (369.2 km^2)
• **Land**	83.87 sq mi (217.2 km^2)
• **Water**	58.67 sq mi (152 km^2)
• **Metro**	8,186 sq mi (21,202 km^2)
Elevation	0–520 ft (0–158 m)
Population (April 1, 2010)[1] [2] [3] [4]	
• **City**	608,660 (US: 23rd)
• **Density**	7361/sq mi (2842.1/km^2)
• **Urban**	2,712,205
• **Metro**	3,439,809 (US: 15th)
• **Demonym**	Seattleite
Time zone	PST (UTC-8)
• **Summer (DST)**	PDT (UTC-7)
ZIP codes	
Area code(s)	206
FIPS code	53-63000[5]
GNIS feature ID	1512650[6]
Website	www.seattle.gov [7]

Seattle (pronounced /siːˈætɫ/ (listen) *see-**at**-əl* or locally /siːˈærɫ/) is the county seat of King County, Washington. With 608,660 residents as of the 2010 Census, Seattle is the largest city in the Northwestern United States. The

Seattle metropolitan area of about 3.4 million inhabitants is the 15th largest metropolitan area in the country.[8] The city is a major seaport situated on a narrow isthmus between Puget Sound (an arm of the Pacific Ocean) and Lake Washington, about 114 miles (183 km) south of the Canada–United States border.

The Seattle area had been inhabited by Native Americans for at least 4,000 years before the first permanent white settlers.[9] Arthur A. Denny and his group of travelers, subsequently known as the Denny Party, arrived at Alki Point on November 13, 1851. The settlement was renamed "Seattle" in 1853, after Chief Seattle of the local Duwamish and Squamish tribes.

Logging was Seattle's first major industry, but by the late 19th century the city had become a commercial and shipbuilding center as a gateway to Alaska during the Klondike Gold Rush. By 1910, Seattle was one of the 25 largest cities in the country.[10] However, a combination of strikes and the Great Depression severely damaged the city's economy. Growth returned during and after World War II when the local Boeing company established Seattle as a center for aircraft manufacturing. Seattle developed as a technology center in the 1980s. The stream of new software, biotechnology, and internet companies lead to an economic revival, which increased the city's population by 50,000 between 1990 and 2000. More recently, Seattle has become a hub for "green" industry and a model for sustainable development.

Seattle has an impressive musical history. From 1918 to 1951, there were nearly two dozen jazz nightclubs along Jackson Street in the current Chinatown/International District. The jazz scene developed the early careers of Ray Charles, Quincy Jones, Ernestine Anderson and others. Seattle is also the birthplace of rock legend Jimi Hendrix and the rock music style known as "grunge,"[11] which was made famous by local groups Nirvana, Soundgarden, Alice in Chains, and Pearl Jam. In more recent years, Seattle has been known for indie rock music.

History

Founding

The Battle of Seattle (1856)

Archaeological excavations confirm that the Seattle area has been inhabited by humans for at least 4,000 years.[9] By the time the first European settlers arrived in the area, the people (now called the Duwamish Tribe) occupied at least seventeen villages in the areas around Elliott Bay.[12] [13] [14]

The first European to visit the Seattle area was George Vancouver, in May 1792 during his 1791–95 expedition to chart the Pacific Northwest.[15]

In 1851, a large party led by Luther Collins made a location on land at the mouth of the Duwamish River; they formally claimed it on September 14, 1851.[16] Thirteen days later, members of the Collins Party on the way to their claim passed three scouts of the Denny Party.[17] Members of the Denny Party claimed land on Alki Point on September 28, 1851.[18] The rest of the Denny Party set sail from Portland, Oregon and landed on Alki point during a rainstorm on November 13, 1851.[18]

After a difficult winter, most of the Denny Party relocated across Elliott Bay and founded the village of "Dewamps" or "Duwamps" on the site of present day Pioneer Square.[18] Charles Terry and John Low remained at the original landing location and established a village they initially called "New York", but renamed "New York Alki" in April 1853, from a Chinook word meaning, roughly, "by and by" or "someday".[19] For the next few years, New York Alki and Duwamps competed for dominance, but in time Alki was abandoned and its residents moved across the bay to join the rest of the settlers.[20]

First Avenue at Columbia Street, c. 1870

David Swinson ("Doc") Maynard, one of Duwamps's founders, was the primary advocate to rename the village "Seattle" after Chief Sealth ("Seattle") of the Duwamish and Suquamish tribes.[21] [22] [23] The term, "Seattle", appears on official Washington Territory papers dated May 23, 1853, when the first plats for the village were filed. In 1855, nominal land settlements were established. On January 14, 1865, the Legislature of Territorial Washington incorporated the Town of Seattle with a board of trustees managing the city. Two years later, after a petition was filed by most of the leading citizens, the Legislature disincorporated the town. In 1867, a young French Canadian Catholic priest named Francis X. Prefontaine arrived in Seattle and decided to establish a parish there. During 1868–69 he built the church by raising the money at fairs in the Puget Sound area and doing much of the work himself, and in 1869 he opened Seattle's first Catholic church at Third Avenue and Washington Street, on the site where the present-day Prefontaine Building stands. The town of Seattle remained a mere precinct of King County until late 1869 when a new petition was filed and the city was re-incorporated with a Mayor-council government.[18] [24] The corporate seal of the City of Seattle carries the date "1869."

Timber town

Seattle's first streetcar, at the corner of Occidental and Yesler, 1884. All of the buildings visible in this picture were destroyed by fire five years later.

Seattle has a history of boom and bust cycles, as is common to cities near areas of extensive natural and mineral resources. Seattle has risen several times economically, then gone into precipitous decline, but it has typically used those periods to rebuild solid infrastructure.[25]

The first such boom, covering the early years of the city, was fueled by the lumber industry. (During this period the road now known as Yesler Way was nicknamed "Skid Road",[26] after the timber skidding down the hill to Henry Yesler's sawmill. This is considered a possible origin for the term which later entered the wider American lexicon as *Skid Row*.)[25] Like much of the American West, Seattle saw numerous conflicts between labor and management, as well as ethnic tensions that culminated in the anti-Chinese riots of 1885–1886.[27] This violence was caused by unemployed whites who determined to drive the Chinese from Seattle (anti-Chinese riots also occurred in Tacoma). In 1900, Asians were 4.2% of the population.[28] Martial law was declared, and federal troops were brought in to put down the disorder. Nevertheless, the economic success in the Seattle area was so great that when the Great Seattle fire of 1889 destroyed the central business district, a far grander city center rapidly emerged in its place.[29] Finance company Washington Mutual, for example, was founded in the immediate wake of the fire.[30] This boom was followed by the construction of a park system, designed by the Olmsted brothers' landscape architecture firm.[25] However, the Panic of 1893 hit Seattle hard.[31]

Gold Rush, World War I, and the Great Depression

The Alaska-Yukon-Pacific Exposition had just over 3.7 million visitors during its 138-day run[32]

The second and most dramatic boom and bust resulted from the Klondike Gold Rush, which ended the depression that had begun with the Panic of 1893; in a short time, Seattle became a major transportation center. On July 14, 1897, the *S.S. Portland* docked with its famed "ton of gold", and Seattle became the main transport and supply point for the miners in Alaska and the Yukon. Few of those working men found lasting wealth, however; it was Seattle's business of clothing the miners and feeding them salmon that panned out in the long run. Along with Seattle, other cities like Everett, Tacoma, Port Townsend, Bremerton, and Olympia, all in the Puget Sound region, became competitors for exchange, rather than mother-lodes for extraction, of precious metals.[33] The boom lasted well into the early part of the 20th century and funded many new Seattle companies and products. In 1907, 19-year-old James E. Casey borrowed $100 from a friend and founded the American Messenger Company (later UPS). Other Seattle companies founded during this period include Nordstrom and Eddie Bauer.[30] The Gold Rush era culminated in the Alaska-Yukon-Pacific Exposition of 1909, which is largely responsible for the layout of today's University of Washington campus.[34]

Pioneer Square in 1917 featuring the Smith Tower, the Seattle Hotel and to the left the Pioneer Building

A shipbuilding boom in the early part of the 20th century became massive during World War I, making Seattle somewhat of a company town; the subsequent retrenchment led to the Seattle General Strike of 1919, the first general strike in the country.[35] A 1912 city development plan by Virgil Bogue went largely unused. Seattle was mildly prosperous in the 1920s but was particularly hard hit in the Great Depression, experiencing some of the country's harshest labor strife in that era. Violence during the Maritime Strike of 1934 cost Seattle much of its maritime traffic, which was rerouted to the Port of Los Angeles.[36]

Seattle was also the home base of impresario Alexander Pantages who, starting in 1902, opened a number of theaters in the city exhibiting vaudeville acts and silent movies. His activities soon expanded, and the thrifty Greek went on and became one of America's greatest theater and movie tycoons. Between Pantages and his rival John Considine, Seattle was for a while the western United States' vaudeville mecca. B. Marcus Priteca, the Scottish-born and Seattle-based architect, built several theaters for Pantages, including some in Seattle. The theaters he built for Pantages in Seattle have been either demolished or converted to other uses, but many other theaters survive in other cities of the USA, often retaining the *Pantages* name; Seattle's surviving Paramount Theatre, on which he collaborated, was not a Pantages theater.

Post-war years: aircraft and software

War work again brought local prosperity during World War II, this time centered on Boeing aircraft. The war dispersed the city's numerous Japanese-American businessmen due to the Japanese American internment. After war, the local economy dipped. It rose again with Boeing's growing dominance in the commercial airliner market.[37] Seattle celebrated its restored prosperity and made a bid for world recognition with the Century 21 Exposition, the 1962 World's Fair.[38] Another major local economic downturn was in the late 1960s and early 1970s. Many people left the area to look for work elsewhere, and two local real estate agents put up a billboard reading "Will the last person leaving Seattle – Turn out the lights."[39]

Building the Seattle Center Monorail, 1961. Looking up Fifth Avenue from Virginia Street.

Seattle remained the corporate headquarters of Boeing until 2001, when the company separated its headquarters from its major production facilities; the headquarters were moved to Chicago.[40] The Seattle area is still home to Boeing's Renton narrow-body plant (where the 707, 720, 727, and 757 were assembled, and the 737 is assembled today) and Everett wide-body plant (assembly plant for the 747, 767, 777, and 787). The company's credit union for employees, BECU, remains based in the Seattle area, though it is now open to all residents of Washington.

As prosperity began to return in the 1980s, the city was stunned by the Wah Mee massacre in 1983, when 13 people were killed in an illegal gambling club in the International District, Seattle's Chinatown.[41] Beginning with Microsoft's 1979 move from Albuquerque, New Mexico to nearby Bellevue, Washington,[42] Seattle and its suburbs became home to a number of technology companies including Amazon.com, RealNetworks, McCaw Cellular (now part of AT&T Mobility), VoiceStream (now T-Mobile USA), and biomedical corporations such as HeartStream (later purchased by Philips), Heart Technologies (later purchased by Boston Scientific), Physio-Control (later purchased by Medtronic), ZymoGenetics, ICOS (later purchased by Eli Lilly and Company) and Immunex (later purchased by Amgen). This success brought an influx of new citizens with a population increase within city limits of almost 50,000 between 1990 and 2000,[43] and saw Seattle's real estate become some of the most expensive in the country.[44] Many of the Seattle area's tech companies remain relatively strong, but the frenzied dot-com boom years ended in early 2001.[45] [46]

Downtown Seattle and a ferry at the Central Waterfront.

Seattle in this period attracted widespread attention as home to these many companies, but also by hosting the 1990 Goodwill Games[47] and the APEC leaders conference in 1993, as well as through the worldwide popularity of grunge, a sound that had developed in Seattle's independent music scene.[48] Another bid for worldwide attention—hosting the World Trade Organization Ministerial Conference of 1999—garnered visibility, but not in the way its sponsors desired, as related protest activity and police reactions to those protests overshadowed the conference itself.[49] The city was further shaken by the Mardi Gras Riots in 2001, and was literally shaken the following day by the Nisqually Earthquake.[50]

The UK consulting firm Mercer, in a 2009 assessment "conducted to help governments and major companies place employees on international assignments", ranked Seattle 50th worldwide in quality of living; the survey factored in political stability, personal freedom, sanitation, crime, housing, the natural environment, recreation, banking facilities, availability of consumer goods, education, and public services including transportation.[51]

Geography

Panorama of Seattle as seen from the Space Needle: a nearly 360-degree view that includes (from left) Puget Sound, Queen Anne Hill, Lake Union, Capitol Hill, downtown Seattle, Elliott Bay, and West Seattle.

With a total area of 83.9 square miles, Seattle lies in the geographical co-ordinates of 47.37 North latitude and 122.20 West longitude. The topography of Seattle is hilly. Seattle lies on seven hills including Capitol Hill, First Hill, West Seattle, Beacon Hill, Denny Hill and Queen Anne. The Kitsap and the Olympic peninsulas along with the Olympic mountains lie to the west of Puget Sound, while the Cascade Range and Lake Sammarish lie to the east of Lake Washington. The lush green forests and the numerous water bodies have provided livelihood for many hunting and gathering communities.

Topography

Seattle is located between Puget Sound (an inlet of the Pacific Ocean) to the west and Lake Washington to the east. The city's chief harbor, Elliott Bay, is an inlet of Puget Sound. To the west, beyond Puget Sound, are the Kitsap Peninsula and Olympic Mountains on the Olympic Peninsula; to the east, beyond Lake Washington and the eastside suburbs, are Lake Sammamish and the Cascade Range. Lake Washington's waters flow to Puget Sound through the Lake Washington Ship Canal (consisting of two man-made canals, Lake Union, and the Hiram M. Chittenden Locks at Salmon Bay, ending in Shilshole Bay).

Downtown Seattle is bounded by Elliott Bay (lower left), lower Broadway (from upper left to lower right), Yesler Way (lower right), and Denny Way (obscured by clouds).

The sea, rivers, forests, lakes, and fields surrounding Seattle were once rich enough to support one of the world's few sedentary hunter-gatherer societies. The surrounding area lends itself well to sailing, skiing, bicycling, camping, and hiking year-round.[52] [53]

The city itself is hilly, though not uniformly so.[54] Like Rome, the city is said to lie on seven hills;[55] the lists vary, but typically include Capitol Hill, First Hill, West Seattle, Beacon Hill, Queen Anne, Magnolia, and the former Denny Hill. The Wallingford, Mount Baker and Crown Hill neighborhoods are technically located on hills as well. Many of the hilliest areas are near the city center, with Capitol Hill, First Hill, and Beacon Hill collectively constituting something of a ridge along an isthmus between Elliott Bay and Lake Washington.[56] The break in the ridge between First Hill and Beacon Hill is man-made, the result of two of the many regrading projects that reshaped the topography of the city center.[57] The topography of the city center was also changed by the construction of a seawall and the artificial Harbor Island (completed 1909) at the mouth of the city's industrial Duwamish Waterway. The highest point within city limits is at High Point in West Seattle, roughly located near 35th Ave SW and SW Myrtle St. Other notable hills include Crown Hill, View Ridge/Wedgwood/Bryant, Maple Leaf, Phinney Ridge, Mt. Baker Ridge, Highlands/Carkeek/Bitterlake.

North of the city center, Lake Washington Ship Canal connects Puget Sound to Lake Washington. It incorporates four natural bodies of water: Lake Union, Salmon Bay, Portage Bay, and Union Bay.

Due to its location in the Pacific Ring of Fire, Seattle is in a major earthquake zone. On February 28, 2001, the magnitude 6.8 Nisqually earthquake did significant architectural damage, especially in the Pioneer Square area (built on reclaimed land, as are the Industrial District and part of the city center), but caused no fatalities.[58] Other strong quakes occurred on January 26, 1700 (estimated at 9 magnitude), December 14, 1872 (7.3 or 7.4),[58] April 13, 1949 (7.1),[59] and April 29, 1965 (6.5).[60] The 1949 quake caused eight known deaths, all in Seattle;[59] the 1965 quake caused three deaths in Seattle directly, and one more by heart failure.[60] Although the Seattle Fault passes just south of the city center, neither it[61] nor the Cascadia subduction zone has caused an earthquake since the city's founding. The Cascadia subduction zone poses the threat of an earthquake of magnitude 9.0 or greater, capable of seriously damaging the city and collapsing many buildings, especially in zones built on fill.[62]

According to the United States Census Bureau, the city has a total area of 142.5 square miles (369 km^2),[63] 83.9 square miles (217 km^2) of which is land and 58.7 square miles (152 km^2) water (41.16% of the total area).

Climate

Seattle's climate is usually described as Oceanic or Marine west coast, with fairly mild, wet winters and very warm, dry summers. Like much of the Pacific Northwest, according to the Köppen climate classification it falls within a cool, dry-summer subtropical zone (*Csb*), with 'cool'-summer Mediterranean characteristics.[65] Other climate classification systems, such as Trewartha, place it firmly in the Oceanic zone (*Do*).[66]

Downtown Seattle averages 71 clear (sunny) days a year, with most of those days occurring between May and September[64]

Temperature extremes are moderated by adjacent Puget Sound, the greater Pacific Ocean, and Lake Washington. The region is largely denied Pacific storms by the Olympic Mountains and Arctic air by the Cascade Range. Despite being on the margin of the rain shadow of the Olympic Mountains, the city has a misleading reputation for frequent rain.[67] This reputation comes from the frequency of precipitation in the winter. In an average year, more than 0.01 in/0.3 mm of precipitation falls on 150 days. It is cloudy 201 days and partly cloudy 93 days.[64] The location of official weather and climatic records, the Seattle-Tacoma (nicknamed Sea-Tac) International Airport, is located about 12 miles south of downtown and records more cloudy days and fewer partly cloudy days per year.[68]

At 944mm (37.17 in.), in reality, the city receives less precipitation annually than New York City (1201 mm, 47.28 in.), Atlanta (1290 mm, 50.79 in.), Boston (1055 mm, 41.53 in.), Baltimore (1038 mm, 40.87 in.), Portland, Maine (1128 mm, 44.41 in.), Jacksonville, Florida (1304 mm, 51.34 in.), and most cities on the Eastern Seaboard of the U.S. Seattle was also not listed in a study that revealed the 10 rainiest cities in the continental United States. Most of the precipitation falls as drizzle or light rain. Thunderstorms are rare. Seattle reports thunder on just seven days per year[69] For comparison, Fort Myers, Florida reports thunder on 93 days per year Kansas City 52, and New York City 25.

There are occasional downpours. One downpour occurred on December 2–4, 2007, when sustained hurricane-force winds and widespread heavy rainfall associated with a strong "Pineapple Express" event occurred in the greater Puget Sound area and the western parts of Washington and Oregon. People blamed heavy rain and strong winds for several power outages and at least four deaths. Interstate 5 at Chehalis, Washington was flooded and closed for almost two days. Precipitation totals exceeded 14 inches (356 mm) in some areas with winds topping out at 130 mph (209 km/hr) along coastal Oregon. [70] It became the second wettest event in Seattle history when a little over 5 inches (130 mm) of rain fell on Seattle in a 24 hour period. People claim the rain indirectly led to five deaths and widespread flooding and damage.[71]

Early spring, late fall, and winter usually have many days when it does not rain. Winters are cool and wet with average lows in the mid 30s °F (2–4 °C) on winter nights. Colder weather does sometimes occur. Summers are very dry by comparison and warm, with average daytime highs around near 75 °F (24 °C). Hotter weather occurs during some summer days. Seattle's hottest official recorded temperature was 103 °F (39 °C) on July 29, 2009;[72] the coldest recorded temperature was 0 °F (−18 °C) on January 31, 1950.[73] Eastern suburbs of Seattle, such as Bellevue and Issaquah, are typically even hotter when the temperature soars above 80 °F (27 °C), due to their location closer to downslope winds from the Cascade Mountains and further from Puget Sound; on Seattle's recorded hottest day of July 29, 2009, parts of south Bellevue, Renton and Issaquah peaked at 110 °F (43 °C).[74]

Between October and May, Seattle is mostly or partly cloudy six out of every seven days[64]

Eighty miles (130 km) to the west, the Hoh Rain Forest in Olympic National Park on the western flank of the Olympic Mountains receives an annual average rainfall of 142 inches (361 cm). Sixty miles to the south of Seattle, the state capital Olympia, which is out of the rain shadow, receives an annual average rainfall of 52 inches (132 cm).

Seattle typically receives some snowfall on an annual basis but heavy snow is rare. Average annual snowfall, as measured at Sea-Tac Airport, is 8.1 inches (21 cm).[75] Single calendar-day snowfall of six inches or greater has occurred on only 15 days since 1948, and only once since February 17, 1990, when 6.8 inches of snow officially fell at Sea-Tac airport on January 18, 2012. This moderate snow event was officially the 12th snowiest calendar day at the airport since 1948 and snowiest since November 1985.[75] Much of the city of Seattle proper received somewhat lesser snowfall accumulations. Locations to the south of Seattle received more, with Olympia and Chehalis receiving 14 to 18 inches even before noon.[76] Another moderate snow event occurred from December 12–25, 2008, when over one foot of snow fell and stuck on much of the roads, causing widespread difficulties in a city not equipped for clearing snow. Seattle's daily record snowfall is 20 inches (51 cm) on January 13, 1950.[77] The largest snowstorm on record occurred from January 5–9, 1880, with snow drifting to 6 feet (1.8 m) in places at the end of the snow event. From January 31 to February 2, 1916, another heavy snow event occurred with 29 inches (74 cm) of snow on the ground by the time the event was over.[78]

A very sunny and dry climate typically dominates from May to late September. An average of 0.8 inches (20 mm) of rain falls in July and 1.0 inch (25 mm) in August. Summer thunderstorms are rare.[79]

The Puget Sound Convergence Zone is an important feature of Seattle's weather. In the convergence zone, air arriving from the north meets air flowing in from the south. Both streams of air originate over the Pacific Ocean; airflow is split by the Olympic Mountains to Seattle's west, then reunited to the east. When the air currents meet, they are forced upward, resulting in convection.[80] Thunderstorms caused by this activity can occur north and south of town, but Seattle itself rarely receives more weather than occasional thunder and small hail showers. The Hanukkah Eve Wind Storm in December 2006 is an exception that brought heavy rain and winds gusting up to 69 mph (111 km/h), not caused by the Puget Sound Convergence Zone.

One of many exceptions to Seattle's reputation as a damp location occurs in El Niño years, when marine weather systems track as far south as California and little precipitation falls in the Puget Sound area.[81] Since the region's water comes from mountain snow packs during the dry summer months, El Niño winters can not only produce substandard skiing but can result in water rationing and a shortage of hydroelectric power the following summer.[82]

Neighborhoods

Downtown Seattle includes a tightly packed financial district along with residential areas and a panoramic waterfront.

Seattle has grown through a series of annexations of smaller neighboring communities. On May 3, 1891, Magnolia, Wallingford, Green Lake, and the University District (then known as Brooklyn) were annexed.[86] The town of South Seattle was annexed on October 20, 1905.[87] Between January 7 and September 12, 1907, Seattle nearly doubled its land area by annexing six incorporated towns and areas of unincorporated King County, including Southeast Seattle, Ravenna, South Park, Columbia City, Ballard, and West Seattle.[88] Three years later, after having difficulties paying a $10,000 bill from the county, the city of Georgetown merged with Seattle.[89] Finally, on January 4, 1954, the area between N. 85th Street and N. 145th Street was annexed, including the neighborhoods of Pinehurst, Greenwood, Blue Ridge, Crown Hill, Broadview, Bitter Lake, Haller Lake, Maple Leaf, Lake City, View Ridge and Northgate.[90]

Former Seattle mayor Greg Nickels is among those who have called Seattle "a city of neighborhoods",[91] [92] although the boundaries (and even names) of those neighborhoods are often open to dispute. For example, a Department of Neighborhoods spokeswoman reported that her own neighborhood has gone from "the 'CD' (Central District) to 'Madrona' to 'Greater Madison Valley' and now 'Madrona Park'.[92]

Over a dozen Seattle neighborhoods have Neighborhood Service Centers, originally known in 1972 as "Little City Halls"[93] and even more have their own street fair and/or parade during the summer months.[94] The largest of the city's street fairs feature hundreds of craft and food booths and multiple stages with live entertainment, and draw more than 100,000 people over the course of a weekend.[95] In addition, at least half a dozen neighborhoods have weekly farmers' markets, some with as many as fifty vendors.[96]

Additionally, Puget Sound Regional Council designates several areas of Seattle as urban centers, defined as "designated planning districts intended to provide a mix of housing, employment and commercial and cultural amenities in a compact form that supports transit, walking and cycling."[97] These urban centers may have the same name as a neighborhood but slightly different borders; for example, the Capitol Hill Urban Center is much smaller than the entire neighborhood.

Cityscape

Landmarks

Image showing 5th Avenue entrance of the Seattle Central Library, designed by OMA; located on 4th and Madison street in Downtown Seattle. Columbia Center can also be seen in the background.

The Space Needle, dating from the Century 21 Exposition (1962), is Seattle's most recognizable landmark, having been featured in the logo of NBA sports team the *Seattle SuperSonics*, the MLS sports team the "Seattle Sounders", the television show *Frasier* and the backgrounds of the television series *Dark Angel*, *Grey's Anatomy* and *iCarly*, and films such as *It Happened at the World's Fair*, *Sleepless in Seattle*, and *The Twilight Saga: Eclipse*. The fairgrounds surrounding the Needle have been converted into Seattle Center, which remains the site of many local civic and cultural events, such as Bumbershoot, Folklife, and the Bite of Seattle. Seattle Center plays multiple roles in the city, ranging from a public fair ground to a civic center, though recent economic losses have called its viability and future into question.[98] The Seattle Center Monorail was also constructed for Century 21 and still runs from Seattle Center to Westlake Center, a downtown shopping mall, a little over a mile to the southeast.

Pike Place Market

The Smith Tower was the tallest building on the West Coast from its completion in 1914 until the Space Needle overtook it in 1962.[99] The late 1980s saw the construction of Seattle's two tallest skyscrapers: the 76 story Columbia Center (completed 1985) is the tallest building in the Pacific Northwest[100] and the fourth tallest building west of the Mississippi River;[101] the Washington Mutual Tower (completed 1988) is Seattle's second tallest building.[102] [103] Other notable Seattle landmarks include Pike Place Market, the Fremont Troll, the Experience Music Project and Science Fiction Museum and Hall of Fame (at Seattle Center), and the Seattle Central Library.

Starbucks has been at Pike Place Market since the coffee company was founded there in 1971. The first store is still operating a block south of its original location.[104]

The National Register of Historic Places has over 150 Seattle listings.[105] The city also designates its own landmarks.[106]

Culture

Nicknames

From 1869 until 1982, Seattle was known as the "Queen City".[107] Seattle's current official nickname is the "Emerald City", the result of a contest held in 1981;[108] [109] the reference is to the lush evergreen forests of the area. Seattle is also referred to informally as the "Gateway to Alaska", "Rain City",[110] and "Jet City", the last from the local influence of Boeing. Seattle residents are known as *Seattleites*.

Performing arts

Seattle has been a regional center for the performing arts for many years. The century-old Seattle Symphony Orchestra is among the world's most recorded[111] and performs primarily at Benaroya Hall.[112] The Seattle Opera and Pacific Northwest Ballet, which perform at McCaw Hall (opened 2003 on the site of the former Seattle Opera

House at Seattle Center), are comparably distinguished,[113] [114] with the Opera being particularly known for its performances of the works of Richard Wagner[115] [116] and the PNB School (founded in 1974) ranking as one of the top three ballet training institutions in the United States.[113] The Seattle Youth Symphony Orchestras (SYSO) is the largest symphonic youth organization in the United States.[117] The city also boasts lauded summer and winter chamber music festivals organized by the Seattle Chamber Music Society.[118]

The 5th Avenue Theatre, built in 1926, stages Broadway-style musical shows[119] featuring both local talent and international stars.[120] Seattle has "around 100" theatrical production companies[121] and over two dozen live theatre venues, many of them associated with fringe theatre;[122] [123] Seattle is probably second only to New York for number of equity theaters[124] (28 Seattle theater companies have some sort of Actors' Equity contract).[121] In addition, the 900-seat Romanesque Revival Town Hall on First Hill hosts numerous cultural events, especially lectures and recitals.[125]

Seattle is considered the home of grunge music[11] because it was home to artists such as Soundgarden, Nirvana, Alice in Chains, Pearl Jam, and Mudhoney, all of whom reached vast audiences in the early 1990s.[126] The city is also home to such varied musicians as avant-garde jazz musicians Bill Frisell and Wayne Horvitz, hot jazz musician Glenn Crytzer, rapper Sir Mix-a-Lot, smooth jazz saxophonist Kenny G, Heart, heavy metal bands Queensrÿche, Demon Hunter, Metal Church, Nevermore, Himsa, and Sunn O))), as well as such poppier rock bands as Harvey Danger, Goodness, and The Presidents of the United States of America. Such musicians as Jimi Hendrix, Duff McKagan, Nikki Sixx, and Quincy Jones spent their formative years in Seattle.

The Moore Theatre has been a performing arts venue in Downtown Seattle since its construction in 1907.

Since the grunge era, the area has hosted a diverse and influential alternative music scene. The Seattle record label Sub Pop—the first to sign Nirvana and Soundgarden—has signed such non-grunge bands as Band of Horses, Modest Mouse, Murder City Devils, Sunny Day Real Estate, Death Cab for Cutie, The Postal Service, Iron and Wine, Flight of the Conchords, and Fleet Foxes. Seattle has also put out a number of hiphop artists, such as Blue Scholars, Macklemore, Fresh Espresso, and group Shabazz Palaces.[126]

Earlier Seattle-based popular music acts include the collegiate folk group The Brothers Four; The Wailers, a 1960s garage band; The Ventures, an instrumental rock band; pop Young Fresh Fellows and The Posies; pop-punk The Fastbacks; the well-traveled avant-rock of Sun City Girls; and the outright punk of The Fartz (later 10 Minute Warning), The Gits.[127]

Over the years, a number of songs have been written about Seattle.

Seattle annually sends a team of spoken word slammers to the National Poetry Slam and considers itself home to such performance poets as Buddy Wakefield, two-time Individual World Poetry Slam Champ;[128] Anis Mojgani, two-time National Poetry Slam Champ;[129] and Danny Sherrard, 2007 National Poetry Slam Champ and 2008 Individual World Poetry Slam Champ.[130] Seattle also hosted the 2001 national Poetry Slam Tournament. The Seattle Poetry Festival is a biennial poetry festival that (launched first as the Poetry Circus in 1997) has featured local, regional, national, and international names in poetry.[131]

The city also has movie houses showing both Hollywood productions and works by independent filmmakers.[132] Among these, the Seattle Cinerama stands out as one of only three movie theaters in the world still capable of showing three-panel Cinerama films.[133]

Additionally, the city is also home to the Seattle Polish Film Festival, ("SPFF") an annual film festival showcasing current and past films of Polish cinema.[134] [135] The festival is produced by the Seattle-Gdynia Sister City Association and awards the Seattle Spirit of Polish Cinema awards as well as the Viewers Choice of Best Film.

Media

As of 2010, Seattle has one major daily newspaper, *The Seattle Times*. The *Seattle Post-Intelligencer*, known as the *P-I*, published a daily newspaper from 1863 to March 17, 2009 before switching to a strictly on-line publication. There is also the *Seattle Daily Journal of Commerce*,[136] and the University of Washington publishes *The Daily*, a student-run publication, when school is in session. The most prominent weeklies are the *Seattle Weekly* and *The Stranger*; both consider themselves "alternative" papers.[137] *Real Change* is a weekly street newspaper that is sold mainly by homeless persons as an alternative to panhandling. There are also several ethnic newspapers, including the *Northwest Asian Weekly*, and numerous neighborhood newspapers, including the *North Seattle Journal*.

Seattle is also well served by television and radio, with all major U.S. networks represented, along with at least five other English-language stations and two Spanish-language stations.[138] Seattle cable viewers also receive CBUT 2 (CBC) from Vancouver, British Columbia.

Non-commercial radio stations include NPR affiliates KUOW-FM 94.9 and KPLU-FM 88.5 (Tacoma). Other stations include KEXP-FM 90.3 (affiliated with EMP), community radio KBCS-FM 91.3 (affiliated with Bellevue College), and high school radio KNHC-FM 89.5, which broadcasts an electronic music radio format and is owned by the public school system and operated by students of Nathan Hale High School. Many Seattle radio stations are also available through Internet radio, with KEXP in particular being a pioneer of Internet radio.[139] Seattle also has numerous commercial radio stations, including KING-FM, one of the last commercial classical music stations in the United States.[138]

Seattle-based online magazines Worldchanging and Grist.org were two of the "Top Green Websites" in 2007 according to TIME.[140]

Seattle also has many online news media websites. The two largest are *The Seattle Times* and *Seattle Post-Intelligencer*.

Tourism

210 cruise ship visits brought 886,039 passengers to Seattle in 2008.[141]

Among Seattle's prominent annual fairs and festivals are the 24-day Seattle International Film Festival,[142] Northwest Folklife over the Memorial Day weekend, numerous Seafair events throughout July and August (ranging from a Bon Odori celebration to the Seafair Cup hydroplane races), the Bite of Seattle, one of the largest Gay Pride festivals in the United States, and the art and music festival Bumbershoot, which programs music as well as other art and entertainment over the Labor Day weekend. All are typically attended by 100,000 people annually, as are the Seattle Hempfest and two separate Independence Day celebrations.[143] [144] [145] In the past, the Gay Pride parade and festival have been centered on Capitol Hill, but since 2006, festivities have been held city-wide, and the parade has followed a route downtown along 4th Avenue from the central shopping district to Seattle Center.[146]

Other significant events include numerous Native American pow-wows, a Greek Festival hosted by St. Demetrios Greek Orthodox Church in Montlake, and numerous ethnic festivals (many associated with Festál at Seattle Center).[147]

Seattle Sheraton from Two Union Square

The Seattle skyline viewed from Gas Works Park.

There are other annual events, ranging from the Seattle Antiquarian Book Fair & Book Arts Show;[148] an anime convention, Sakura-Con;[149] Penny Arcade Expo, a gaming convention;[150] a two-day, 9,000-rider Seattle to Portland Bicycle Classic,[151] and specialized film festivals, such as the Maelstrom International Fantastic Film Festival, the Northwest Asian-American Film Festival, Children's Film Festival Seattle, Translation: the Seattle Transgender Film Festival, and the Seattle Gay and Lesbian Film Festival.[152] [153]

The Henry Art Gallery opened in 1927, the first public art museum in Washington.[154] The Seattle Art Museum (SAM) opened in 1933; SAM opened a museum downtown in 1991 (expanded and reopened 2007); since 1991, the 1933 building has been SAM's Seattle Asian Art Museum (SAAM).[155] SAM also operates the Olympic Sculpture Park (opened 2007) on the waterfront north of the downtown piers. The Frye Art Museum is a free museum on First Hill. Regional history collections are at the Loghouse Museum in Alki, Klondike Gold Rush National Historical Park, the Museum of History and Industry and the Burke Museum of Natural History and Culture. Industry collections are at the Center for Wooden Boats and the adjacent Northwest Seaport, the Seattle Metropolitan Police Museum, and the Museum of Flight. Regional ethnic collections include the Nordic Heritage Museum, the Wing Luke Asian Museum and the Northwest African American Museum. Seattle has artist-run galleries,[156] including 10-year veteran Soil Art Gallery,[157] and the newer Crawl Space Gallery.[158]

Woodland Park Zoo opened as a private menagerie in 1889, but was sold to the city in 1899.[159] The Seattle Aquarium has been open on the downtown waterfront since 1977 (undergoing a renovation 2006).[160] The Seattle Underground Tour is an exhibit of places that existed before the Great Fire.[161] There are also many community centers for recreation, including Rainier Beach, Van Asselt, Rainier, and Jefferson south of the Ship Canal and Green Lake, Laurelhurst, Loyal Heights north of the Canal, and Meadowbrook.[162]

Since the middle 1990s, Seattle has experienced significant growth in the cruise industry, especially as a departure point for Alaska cruises. In 2008, a record total of 886,039 cruise passengers passed through the city, surpassing the number for Vancouver, BC, the other major departure point for Alaska cruises.[163]

Professional sports

Safeco Field, home of the Seattle Mariners

CenturyLink Field, home of the Seattle Seahawks and Seattle Sounders FC

Club	Sport	League	Venue	Established	Championships
Seattle Grizzlies	Australian Football	USAFL	Mosier Park	1998	0
Seattle Mist	Football	LFL	ShoWare Center	2009	N/AQ
Seattle Mariners	Baseball	MLB	Safeco Field	1977	0
Seattle Seahawks	Football	NFL	CenturyLink Field	1976	0
Seattle Sounders FC	Soccer	MLS	CenturyLink Field	2007	2[164]
Seattle Storm	Basketball	WNBA	KeyArena	2000	2
Seattle SuperSonics*	Basketball	NBA	KeyArena	1967	1
Seattle Thunderbirds	Ice hockey	WHL	ShoWare Center	1977	0
Seattle Totems	Ice Hockey	NPHL	Olympic View Ice Arena	2005	4

Seattle's professional sports history began at the start of the 20th century with the PCHA's Seattle Metropolitans, which in 1917 became the first American hockey team to win the Stanley Cup.[165] Today Seattle has three major professional sports teams: The National Football League's Seattle Seahawks, Major League Baseball's Seattle Mariners, and Major League Soccer's Seattle Sounders FC. Other sports teams include the 2004 and 2010 Women's National Basketball Association champions, Seattle Storm.[166] From 1967 to 2008 Seattle was also home to an NBA franchise, the Seattle SuperSonics, who were the 1978–79 NBA champions. The team relocated to Oklahoma City after the 2007–08 season.[167] The Seattle Thunderbirds are a major-junior hockey team that plays in one of the Canadian major-junior hockey leagues, the WHL (Western Hockey League). The Thunderbirds moved to nearby Kent, Washington during the 2008–2009 season.[168] The Seattle Sounders FC began play in Major League Soccer in 2009.[169]

The Major League Baseball All-Star game was held in Seattle twice, first at the Kingdome in 1979 and again at Safeco Field in 2001. That same year, the Seattle Mariners tied the all-time single regular season wins record with 116 wins. The NBA All-Star game was also held in Seattle twice, the first in 1974 at the Seattle Center Coliseum and the second in 1987 at the Kingdome.[170]

In 2006, CenturyLink Field hosted the 2005–06 NFL playoffs. In 2008, CenturyLink Field hosted the first game of the 2007–08 NFL playoffs, in which the Seahawks defeated the Washington Redskins, 35–14. CenturyLink Field also serves as the home field for the Seattle Sounders FC of Major League Soccer. Seattle also boasts a strong history in collegiate sports. The University of Washington and Seattle University are NCAA Division I schools.

Outdoor activities

Green Lake Park, popular among runners, contains a 2.7-mile (4.3 km) trail circling the lake.

Seattle's mild, temperate marine climate allows year-round outdoor recreation, including walking, cycling, hiking, skiing, snowboarding, kayaking, rock climbing, motor boating, sailing, team sports, and swimming.[171] In town, many people walk around Green Lake, through the forests and along the bluffs and beaches of 535-acre (2.2 km^2) Discovery Park (the largest park in the city) in Magnolia, along the shores of Myrtle Edwards Park on the Downtown waterfront, along the shoreline of Lake Washington at Seward Park, along Alki Beach in West Seattle, or along the Burke-Gilman Trail. Also popular are hikes and skiing in the nearby Cascade or Olympic Mountains and kayaking and sailing in the waters of Puget Sound, the Strait of Juan de Fuca, and the Strait of Georgia. In 2005, *Men's Fitness* magazine named Seattle the fittest city in the United States.[172]

Economy

Washington Mutual's last headquarters, the WaMu Center, (now the Chase Center and soon to be Russell Investments Center) (center left) and its headquarters prior, Washington Mutual Tower (center right).

Seattle's economy is driven by a mix of older industrial companies, and "new economy" Internet and technology companies, service, design and clean technology companies. The city's gross metropolitan product was $231 billion in 2010, making it the 12th largest metropolitan economy in the United States.[173] The Port of Seattle, which also operates Seattle–Tacoma International Airport, is a major gateway for trade with Asia and cruises to Alaska, and is the 8th largest port in the United States in terms of container capacity.[174] Though it has been affected by the recent recession, Seattle has retained a comparatively strong economy, and remains a hotbed for start-up businesses, especially in green building and clean technologies: it was ranked as America's No. 1 "smarter city" based on its government policies and green economy.[175] February 2010, the city government committed Seattle to becoming North America's first "climate neutral" city, with a goal of reaching zero net per capita greenhouse gas emissions by 2030.[176]

Still, very large companies dominate the business landscape. Three companies on the 2008 Fortune 500 list of the United States' largest companies, based on total revenue are headquartered in Seattle: Internet retailer Amazon.com (#100), coffee chain Starbucks (#241), and department store Nordstrom (#270).[177] Other Fortune 500 companies popularly associated with Seattle are based in nearby Puget Sound cities. Warehouse club chain Costco (#29), the largest retail company in Washington, is based in Issaquah. Providence Health & Services, the largest health care system and the fifth largest employer, is based in Renton, Washington. Microsoft (#44) and Nintendo of America are located in Redmond. Weyerhaeuser, the forest products company (#147), is based in Federal Way. Finally, Bellevue is home to truck manufacturer PACCAR (#169), and to international mobile telephony giant T-Mobile's U.S. subsidiary, T-Mobile USA.[177] The city has a reputation for heavy coffee consumption;[178] coffee companies founded or based in Seattle include Starbucks,[179] Seattle's Best Coffee,[180] [181] and Tully's.[182] There are also many successful independent artisanal espresso roasters and cafes.[178]

Prior to moving its headquarters to Chicago, aerospace manufacturer Boeing (#27) was the largest company based in Seattle. Its largest division is still headquartered in nearby Renton, and the company has large aircraft manufacturing plants in Everett and Renton, so it remains the largest private employer in the Seattle metropolitan area.[183] Seattle Mayor Greg Nickels announced a desire to spark a new economic boom driven by the biotechnology industry in

2006. Major redevelopment of the South Lake Union neighborhood is underway, in an effort to attract new and established biotech companies to the city, joining biotech companies Corixa (acquired by GlaxoSmithKline), Immunex (now part of Amgen), Trubion, and ZymoGenetics. Vulcan Inc., the holding company of billionaire Paul Allen, is behind most of the development projects in the region. While some see the new development as an economic boon, others have criticized Nickels and the Seattle City Council for pandering to Allen's interests at taxpayers' expense.[184] Also in 2006, *Expansion Magazine* ranked Seattle among the top 10 metropolitan areas in the nation for climates favorable to business expansion.[185] In 2005, *Forbes* ranked Seattle as the most expensive American city for buying a house based on the local income levels.[186]

Alaska Airlines, operating a hub at Seattle–Tacoma International Airport, maintains its headquarters in the city of SeaTac, next to the airport.[187]

Demographics

Historical populations		
Census	**Pop.**	**%±**
1860	188	—
1870	1151	512.2%
1880	3533	207.0%
1890	42837	1112.5%
1900	80671	88.3%
1910	237194	194.0%
1920	315312	32.9%
1930	365583	15.9%
1940	368302	0.7%
1950	467591	27.0%
1960	557087	19.1%
1970	530831	−4.7%
1980	493846	−7.0%
1990	516259	4.5%
2000	563374	9.1%
2010	608660	8.0%
source:[188] [189]		

According to the 2010 census, Seattle had a population of 608,660 and the racial composition was as follows:[190]

- White: 69.5% (Non-Hispanic Whites: 66.3%)
- Asian: 13.8% (4.1% Chinese, 2.6% Filipino, 2.2% Vietnamese, 1.3% Japanese, 1.1% Korean, 0.8% Indian, 0.3% Cambodian, 0.3% Laotian, 0.2% Thai)
- Black or African American: 7.9%
- American Indian and Alaska Native: 0.8%
- Native Hawaiian and Other Pacific Islander: 0.4%
- Other Race: 2.4%
- Two or more races: 5.1%

- Hispanic or Latino (of any race): 6.6% (4.1% Mexican, 0.3% Puerto Rican, 0.2% Guatemalan, 0.2% Salvadoran, 0.2% Cuban)

Seattle's population has historically been predominantly white.[28] In 1960, whites comprised 91.6% of the city's population.[28]

According to the 2006–2008 American Community Survey, English was by far the most commonly spoken language at home; approximately 78.9% of residents over the age of five spoke only English at home. Spanish was spoken by 4.5% of the population; people who spoke other Indo-European languages made up 3.9% of the population. People who spoke Asian languages at home made up 10.2% of the population. People who spoke other languages made up 2.5% of Seattle's population.[191]

Seattle has seen a major increase in immigration in recent decades; the foreign-born population increased 40% between the 1990 and 2000 censuses.[192] The Chinese population in the Seattle area has origins in mainland China, Hong Kong, Southeast Asia, and Taiwan. The earliest Chinese Americans that came in the late 19th and early 20th century were almost entirely from Guangdong province. The Seattle area is also home to a high Vietnamese population.[193] In addition, the Seattle area is home to over 30,000 Somali immigrants.[194]

As of 1999, the median income of a city household was $45,736, and the median income for a family was $62,195. Males had a median income of $40,929 versus $35,134 for females. The per capita income for the city was $30,306.[195] 11.8% of the population and 6.9% of families are below the poverty line. Of people living in poverty, 13.8% are under the age of 18 and 10.2% are 65 or older.[195]

It is estimated that King County has 8,000 homeless people on any given night, and many of those live in Seattle.[196] In September 2005, King County adopted a "Ten-Year Plan to End Homelessness", one of the near-term results of which is a shift of funding from homeless shelter beds to permanent housing.[197]

In 2006, after growing by 4,000 citizens per year for the previous 16 years, regional planners expected the population of Seattle to grow by 200,000 people by 2040.[198] However, Mayor Nickels supported plans that would increase the population by 60%, or 350,000 people, by 2040 and is working on ways to accommodate this growth while keeping Seattle's single-family housing zoning laws.[198] The Seattle City Council later voted to relax height limits on buildings in the greater part of Downtown, partly with the aim to increase residential density in the city center.[199]

A 2006 study by UCLA indicated that Seattle has one of the highest LGBT populations per capita. With 12.9% of citizens polled identifying as gay, lesbian, or bisexual, the city ranks second of all major US cities, behind San Francisco and slightly ahead of Atlanta and Minneapolis.[200] Greater Seattle also ranks second among major US metropolitan areas, with 6.5% being LGBT.[201]

According to the 2000 US census interim measurements of 2004, Seattle has the fifth highest proportion of single-person households nationwide among cities of 100,000 or more residents, at 40.8%.[202]

Government and politics

Seattle City Hall, 2007

Seattle is a charter city, with a Mayor–Council form of government. Since 1911, Seattle's nine city councillors have been elected at large, rather than by geographic subdivisions.[203] The only other elected offices are the city attorney and Municipal Court judges. All city, county, and state offices are technically non-partisan.[204] Like most parts of the United States, government and laws are also run by a series of ballot initiatives (where people can pass or reject laws), referendums (where people can approve or reject already passed legislation), and Propositions (where specific government agencies can propose new laws/tax increases directly to the people)

Seattle's politics are strongly liberal/progressive, although there is a small libertarian movement within the metro area.[205] It is one of the most liberal cities in the United States, with approximately 80% voting for the Democratic Party; only two precincts in Seattle—one in the Broadmoor community, and one encompassing neighboring Madison Park—had a majority of votes for Republican George W. Bush in the 2004 presidential election. In addition, all precincts in Seattle voted for Democratic Party candidate Barack Obama in the 2008 presidential election, including the two precincts who had previously voted Republican in 2004.[205] In partisan elections for the Washington State Legislature and United States Congress, nearly all elections are won by Democrats.

Seattle is one of the most politically progressive cities in North America, with an overwhelming majority of voters supporting Democratic politicians; support for liberal issues such as same-sex marriage, reproductive rights and gun control is largely taken for granted in local politics. Like much of the Pacific Northwest (which has the lowest rate of church attendance in the United States and consistently reports the highest percentage of atheism[206] [207]), church attendance, religious belief and political influence of religious leaders is much lower than in other parts of America.[208] Seattle also has a thriving alternative press, with two well-established weekly newspapers, several online dailies (including the *Seattle P.I.*, *Publicola* and *Crosscut*), and a number of issue-focused publications, including the nation's two largest online environmental magazines, *Worldchanging* and *Grist.org*.

Federally, Seattle is part of Washington's 7th congressional district, represented by Democrat Jim McDermott, elected in 1988 and one of Congress' most liberal members.[209]

Education

Of the city's population over the age of 25, 53.8% (vs. a national average of 27.4%) hold a bachelor's degree or higher, and 91.9% (vs. 84.5% nationally) have a high school diploma or equivalent.[210] A United States Census Bureau survey showed that Seattle had the highest percentage of college and university graduates of any major U.S. city.[211] The city was listed as the most literate of the country's 69 largest cities in 2005 and 2006 and the second most literate in 2007, after Minneapolis, and tied with Minneapolis for most literate in 2008 in studies conducted by Central Connecticut State University.[212]

Seattle Public Schools desegregated without a court order[213] but continue to struggle to achieve racial balance in a somewhat ethnically divided city (the south part of town having more ethnic minorities than the north).[214] In 2007, Seattle's racial tie-breaking system was struck down by the United States Supreme Court, but the ruling left the door open for desegregation formulae based on other indicators (e.g., income or socioeconomic class).[215]

University of Washington Quad in Spring

The public school system is supplemented by a moderate number of private schools: five of the private high schools are Catholic, one is Lutheran, and six are secular.[216]

Seattle is home to one of the United States's most respected public research universities, the University of Washington, as well as its professional and continuing education unit, University of Washington Educational Outreach. A study by *Newsweek International* in 2006 cited UW as the twenty-second best university in the world.[217] Seattle also has a number of smaller private universities including Seattle University and Seattle Pacific University, both founded by religious groups; universities aimed at the working adult, like City University and Antioch University; colleges, such as North Seattle Community College, Seattle Central Community College, and South Seattle Community College; and a number of arts colleges, such as Cornish College of the Arts and The Art Institute of Seattle. In 2001, *Time* magazine selected Seattle Central Community College as community college of the year, stating the school "pushes diverse students to work together in small teams".[218]

Infrastructure

Health systems

The University of Washington is consistently ranked among the country's top leading institutions in medical research, earning special merits for programs in neurology and neurosurgery. Seattle has seen local developments of modern paramedic services with the establishment of Medic One in 1970.[219] In 1974, a *60 Minutes* story on the success of the then four-year-old Medic One paramedic system called Seattle "the best place in the world to have a heart attack".[220]

Three of Seattle's largest medical centers are located on First Hill. Harborview Medical Center, the public county hospital, is the only Level I trauma hospital serving Washington, Alaska, Montana, and Idaho.[221] Virginia Mason Medical Center and Swedish Medical Center's two largest campuses are also located in this part of Seattle, including the Virginia Mason Hospital. This concentration of hospitals resulted in the neighborhood's nickname "Pill Hill".[222]

Located in the Laurelhurst neighborhood, Seattle Children's, formerly Children's Hospital and Regional Medical Center, is the pediatric referral center for Washington, Alaska, Montana, and Idaho. The Fred Hutchinson Cancer Research Center has a campus in the Eastlake neighborhood and also shares facilities with the Seattle Cancer Care Alliance and University of Washington Medical Center. The University District is home to the University of Washington Medical Center which, along with Harborview, is operated by the University of Washington. Seattle is also served by a Veterans Affairs hospital on Beacon Hill, a third campus of Swedish in Ballard, and Northwest Hospital and Medical Center near Northgate Mall.

Transportation

Further information: Street layout of Seattle

Interstate 5 in Washington as it passes through downtown Seattle

The first streetcars appeared in 1889 and were instrumental in the creation of a relatively well-defined downtown and strong neighborhoods at the end of their lines. The advent of the automobile sounded the death knell for rail in Seattle. Tacoma–Seattle railway service ended in 1929 and the Everett–Seattle service came to an end in 1939, replaced by inexpensive automobiles running on the recently developed highway system. Rails on city streets were paved over or removed, and the opening of the Seattle trolleybus system brought the end of streetcars in Seattle in 1941. This left an extensive network of privately owned buses (later public) as the only mass transit within the city and throughout the region.[223]

Central Link light rail trains in the Downtown Seattle Transit Tunnel under the International District/Chinatown.

King County Metro provides frequent stop bus service within the city and surrounding county, as well as a South Lake Union Streetcar line between the South Lake Union neighborhood and Westlake Center in downtown.[224] Seattle is one of the few cities in North America whose bus fleet includes electric trolleybuses. Sound Transit currently provides an express bus service within the metropolitan area; two Sounder commuter rail lines between the suburbs and downtown; and its Central Link light rail line, which opened in 2009, between downtown and Sea-Tac Airport gives the city its first rapid transit line that has intermediate stops within the city limits. Washington State Ferries, which manages the largest network of ferries in the United States and third largest in the world,[225] connects Seattle to Bainbridge and Vashon Islands in Puget Sound and to Bremerton and Southworth on the Kitsap Peninsula.[225]

According to the 2007 American Community Survey, 18.6% of Seattle residents used one of the three public transit systems that serve the city, giving it the highest transit ridership of all major cities without heavy or light rail prior to the completion of Sound Transit's Central Link line.[226] [227] The city has also been described as the fourth most walkable city in the United States.[228]

Seattle–Tacoma International Airport, locally known as Sea-Tac Airport and located just south in the neighboring city of SeaTac, is operated by the Port of Seattle and provides commercial air service to destinations throughout the world. Closer to downtown, Boeing Field is used for general aviation, cargo flights, and testing/delivery of Boeing airliners.

The main mode of transportation, however, relies on Seattle's streets, which are laid out in a cardinal directions grid pattern, except in the central business district where early city leaders Arthur Denny and Carson Boren insisted on orienting their plats relative to the shoreline rather than to true North.[229] Only two roads, Interstate 5 and State Route 99 (both limited-access highways), run uninterrupted through the city from north to south. State Route 99 runs through downtown Seattle on the Alaskan Way Viaduct, which was built in 1953. However, due to damage sustained during the 2001 Nisqually earthquake the viaduct will be replaced by a tunnel in 2015 at a cost of US$4.25 billion.

From 2006 to 2008, transit ridership in Seattle went up by 23%, and many bus routes in the central part of the city are routinely forced to leave passengers because they are full. Seattle has the 12th worst traffic congestion of all American cities.[230]

The city has started moving away from the automobile and towards mass transit. In 2006, voters in King County passed proposition 2(Transit Now) which increased bus service hours on high ridership routes and paid for five Bus Rapid Transit lines called RapidRide.[231] After rejecting a roads and transit measure in 2007, Seattle-area voters passed a transit only measure in 2008 that increases ST Express bus service and extends the Link Light Rail system (currently 15.7 miles (25.3 km) with 3 miles (4.8 km) under construction) by over 30 miles, and expands and improves Sounder commuter rail service.[232] A light rail line from downtown heading south to the Seattle–Tacoma International Airport began service on December 19, 2009, giving the city its first rapid transit line with intermediate stations within the city limits. An extension north to the University of Washington is under construction as of 2010; and further extensions are planned to reach Lynnwood to the north, Des Moines to the south, and Bellevue and Redmond to the east by 2023.[233] [234] Mayor Michael McGinn hopes to put another transit measure on the 2011 ballot to build light rail from Downtown Seattle to Ballard, Fremont, and West Seattle[235] after seeing a surprisingly large amount of support for it from its campaign (and now city's) policy forum.

Walkability

A 2011 study by Walk Score ranked Seattle the sixth most walkable of the fifty largest U.S. cities.[236]

Utilities

Water and electric power are municipal services, provided by Seattle Public Utilities and Seattle City Light respectively. Other utility companies serving Seattle include Puget Sound Energy (natural gas, electricity); Seattle Steam Company (steam); Waste Management, Inc and CleanScapes, Inc. (curbside recycling and solid waste removal); and Verizon Communications, Century Link and Comcast (telephone, Internet, and cable television).

Seattle Steam Company, one of Seattle's privately owned utility companies

Sister cities

- Beer Sheva

- Bergen
- Cebu
- Chongqing
- Christchurch
- Daejeon
- Galway
- Gdynia
- Haiphong
- Kobe
- Limbe
- Mazatlán
- Mombasa
- Nantes
- Pécs
- Perugia
- Reykjavík
- Sihanoukville
- Surabaya
- Tashkent

[237]

See also

- List of people from Seattle
- List of Seattle sister cities
- National Register of Historic Places listings in Seattle, Washington
- Sustainable Seattle
- Tillicum Village
- Seattle tugboats

References

[1] "Latest estimate: Seattle has 608,660 people" (http://www.bizjournals.com/seattle/stories/2010/06/21/daily9.html?ana=e_du_pap). Puget Sound Business Journal. June 22, 2010. . Retrieved June 22, 2010.

[2] "Alphabetically sorted list of Census 2000 Urbanized Areas" (http://www.census.gov/geo/www/ua/ua2k.txt) (TXT). United States Census Bureau, Geography Division. . Retrieved July 12, 2009.

[3] "Annual Estimates of the Population of Metropolitan and Micropolitan Statistical Areas: April 1, 2000 to July 1, 2008 (CBSA-EST2008-01)" (http://www.census.gov/popest/metro/tables/2008/CBSA-EST2008-01.csv) (CSV). United States Census Bureau, Population Division. . Retrieved July 12, 2009.

[4] "Demographia World Urban Areas & Population Projections" (http://www.demographia.com/db-worldua.pdf) (PDF). Demographia. April 2009. pp. 35, 91. . Retrieved April 18, 2009.

[5] "American FactFinder" (http://factfinder.census.gov). United States Census Bureau. . Retrieved 2008-01-31.

[6] "US Board on Geographic Names" (http://geonames.usgs.gov). United States Geological Survey. 2007-10-25. . Retrieved 2008-01-31.

[7] http://www.seattle.gov/

[8] "Estimates of Population Change for Metropolitan Statistical Areas and Rankings: July 1, 2007 to July 1, 2008 (CBSA-EST2008-05)" (http://www.census.gov/popest/metro/tables/2008/CBSA-EST2008-05.csv) (CSV). United States Census Bureau, Population Division. . Retrieved July 12, 2009.

[9] Doree Armstrong (October 4, 2007). "Feel the beat of history in the park and concert hall at two family-friendly events" (http://www.seattlepi.com/lifestyle/article/Feel-the-beat-of-history-in-the-park-and-concert-1251579.php). Seattle Post-Intelligencer. . Retrieved November 1, 2007.

[10] "Population of the 100 Largest Urban Places: 1910" (http://www.census.gov/population/www/documentation/twps0027/tab14.txt). U.S. Bureau of the Census. . Retrieved 16 November 2011.

[11] Heylin, Clinton (2007). *Babylon's Burning: From Punk to Grunge*. Conongate. p. 606. ISBN 978-1-84195-879-8.

[12] Greg Lange (October 15, 2000). "Seattle and King County's First White Settlers" (http://www.historylink.org/index.cfm?DisplayPage=output.cfm&File_Id=1660). HistoryLink. . Retrieved October 14, 2007.

[13] "The people and their land" (http://www.seattleartmuseum.org/learn/CDROM/SongStorySpeech/Content/SalishArtCulture.htm). *Puget Sound Native Art and Culture*. Seattle Art Museum. c. July 4, 2003 per "Native Art of the Northwest Coast: Collection Insight" (http://www.seattleartmuseum.org/exhibit/exhibitDetail.asp?WHEN=PAST&eventID=2926). . Retrieved April 21, 2006.

[14] Crowley, Walt (March 13, 2003). "Native American tribes sign Point Elliott Treaty at Mukilteo on January 22, 1855." (http://www.historylink.org/index.cfm?DisplayPage=output.cfm&File_Id=5402). HistoryLink.org Essay 5402. . Retrieved October 14, 2007.

[15] Vancouver, George, and John Vancouver (1801). *A voyage of discovery to the North Pacific ocean, and round the world* (http://books.google.com/?id=qwol8bPaYxsC&printsec=frontcover). London: J. Stockdale. ISBN 978-0-665-18642-4. .

[16] Greg Lange (March 8, 2003). "Luther Collins Party, first King County settlers, arrive at mouth of Duwamish River on September 14, 1851." (http://www.historylink.org/index.cfm?DisplayPage=output.cfm&File_Id=5390). HistoryLink. . Retrieved October 14, 2007.

[17] Greg Lange (December 16, 2000). "Collins party encounters Denny party scouts at Duwamish Head near future site of Seattle on September 27, 1851." (http://www.historylink.org/index.cfm?DisplayPage=output.cfm&File_Id=2765). HistoryLink. . Retrieved October 14, 2007.

[18] Crowley, Walt (August 31, 1998). "Seattle – a Snapshot History of Its Founding" (http://www.historylink.org/index.cfm?DisplayPage=output.cfm&File_Id=303). HistoryLink. . Retrieved October 14, 2007.

[19] James R. Warren (October 23, 2001). "Seattle at 150: Charles Terry's unlimited energy influenced a city" (http://www.seattlepi.com/local/article/Seattle-at-150-Charles-Terry-s-unlimited-energy-1069610.php). Seattle Post-Intelligencer. . Retrieved October 14, 2007.

[20] Greg Lange (March 28, 2001). "Charles Terry homesteads site of Alki business district on May 1, 1852." (http://www.historylink.org/index.cfm?DisplayPage=output.cfm&File_Id=3142). HistoryLink. . Retrieved October 14, 2007.

[21] Thomas R. Speer, editor, ed (July 22, 2004). "Chief Si'ahl and His Family" (http://www.duwamishtribe.org/chiefsiahl.html). Duwamish Tribe. . Retrieved October 14, 2007., includes bibliography.

[22] Kenneth G. Watson (January 18, 2003). "Seattle, Chief Noah" (http://www.historylink.org/index.cfm?DisplayPage=output.cfm&File_Id=5071). HistoryLink. . Retrieved October 14, 2007.

[23] Morgan, Murray (1982 (originally published 1951, 1982 revised and updated, first illustrated edition)). *Skid Road: an Informal Portrait of Seattle*. Seattle and London: University of Washington Press. p. 20. ISBN 978-0-295-95846-0.

[24] Greg Lange; Cassandra Tate (November 4, 1998). "Legislature incorporates the Town of Seattle for the first time on January 14, 1865." (http://www.historylink.org/index.cfm?DisplayPage=output.cfm&File_Id=168). HistoryLink. . Retrieved October 14, 2007.

[25] Emmett Shear (Spring 2002). *Seattle: Booms and Busts*. Yale University. Author has granted blanket permission for material from that paper to be reused in Wikipedia. Now at s:Seattle: Booms and Busts.

[26] Junius Rochester (October 7, 1998). "Yesler, Henry L." (http://www.historylink.org/index.cfm?DisplayPage=output.cfm&File_Id=286). HistoryLink. . Retrieved October 1, 2007.

[27] George Kinnear (January 1, 1911). "Anti-Chinese Riots At Seattle, Wn.. February 8, 1876" (http://en.wikisource.org/w/index.php?title=Anti-Chinese_Riots_At_Seattle). *Seattle Post-Intelligencer*. . Retrieved October 4, 2007. Kinnear's article originally appeared in the *Seattle Post-Intelligencer* and was later privately published in a small volume.

[28] "Historical Census Statistics On Population Totals By Race, 1790 to 1990, and By Hispanic Origin, 1970 to 1990, For Large Cities And Other Urban Places In The United States" (http://www.census.gov/population/www/documentation/twps0076/twps0076.html). U.S. Census Bureau. . Retrieved 2011-12-18.

[29] Walt Crowley (January 25, 2003). "Seattle burns down in the Great Fire on June 6, 1889." (http://www.historylink.org/index.cfm?DisplayPage=output.cfm&File_Id=5115). HistoryLink. . Retrieved October 1, 2007.

[30] "Hard Drive to the Klondike: Promoting Seattle During the Gold Rush" (http://web.archive.org/web/20071103062325/http://www.nps.gov/klse/hrs/hrs0.htm). National Park Service. February 18, 2003. Archived from the original (http://www.nps.gov/klse/hrs/hrs0.htm) on November 3, 2007. . Retrieved October 1, 2007.

[31] J. Kingston Pierce (November 24, 1999). "Panic of 1893: Seattle's First Great Depression." (http://www.historylink.org/index.cfm?DisplayPage=output.cfm&File_Id=2030). HistoryLink. . Retrieved December 18, 2008.

[32] Greg Lange (January 14, 1999). "Alaska–Yukon–Pacific Exposition's final day is on October 16, 1909." (http://www.historylink.org/index.cfm?DisplayPage=output.cfm&File_Id=696). HistoryLink. . Retrieved November 6, 2007.

[33] Greg Lange (January 14, 1999). "Klondike Gold Rush" (http://www.historylink.org/index.cfm?DisplayPage=output.cfm&File_Id=687). HistoryLink.org. . Retrieved October 1, 2007.

[34] Greg Lange (May 5, 2003). "Alaska–Yukon–Pacific Exposition opens for a 138-day run on June 1, 1909." (http://www.historylink.org/index.cfm?DisplayPage=output.cfm&File_Id=5371). HistoryLink. . Retrieved October 1, 2007.

[35] Patrick McRoberts (February 4, 1999). "Seattle General Strike, 1919, Part I" (http://www.historylink.org/index.cfm?DisplayPage=output.cfm&File_Id=861). HistoryLink. . Retrieved October 1, 2007.

[36] BOLA Architecture + Planning & Northwest Archaeological Associates, Inc., Port of Seattle North Bay Project DEIS: Historic and Cultural Resources (http://web.archive.org/web/20110723025013/http://www.portseattle.org/downloads/business/realestate/development/northbay/Appendix_I_Historic_Cultural.pdf), Port of Seattle, April 5, 2005, p. 12-13 (which is p. 14-15 of the PDF). Retrieved July 25, 2008.

[37] "History of Seattle: The "Jet City" Takes Off" (http://web.archive.org/web/20061002171554/http://www.seeseattle.org/media/presskits/PKHistory.asp). Seattle's Convention and Visitors Bureau. Archived from the original (http://www.seeseattle.org/media/

presskits/PKHistory.asp) on October 2, 2006. . Retrieved October 1, 2007.
[38] Alan J. Stein (April 18, 2000). "Century 21 – The 1962 Seattle World's Fair, Part I" (http://www.historylink.org/index.cfm?DisplayPage=output.cfm&File_Id=2290). HistoryLink. . Retrieved October 1, 2007.
[39] Greg Lange (June 8, 1999). "Billboard appears on April 16, 1971, near Sea–Tac, reading: Will the Last Person Leaving Seattle—Turn Out the Lights." (http://www.historylink.org/index.cfm?DisplayPage=output.cfm&File_Id=1287). HistoryLink. . Retrieved October 1, 2007. The real estate agents were Bob McDonald and Jim Youngren, as cited at Don Duncan, *Washington: the First One Hundred Years, 1889–1989* (Seattle: The Seattle Times, 1989), 108, 109–110; *The Seattle Times*, February 25, 1986, p. A3; Ronald R. Boyce, *Seattle–Tacoma and the Southern Sound* (Bozeman, Montana: Northwest Panorama Publishing, 1986), 99; Walt Crowley, *Rites of Passage: A Memoir of the Sixties in Seattle* (Seattle: University of Washington Press, 1995), 297.
[40] Kristi Heim (March 21, 2006). "Chicago's got the headquarters, but Seattle's still Jet City, USA" (http://seattletimes.nwsource.com/html/businesstechnology/2002876673_boeingimage20.html). The Seattle Times. . Retrieved October 1, 2007.
[41] Natalie Singer (September 7, 2006). "23 years haven't erased grief caused by Wah Mee Massacre" (http://seattletimes.nwsource.com/html/localnews/2003247239_wahmee07m.html). Seattle Times. . Retrieved December 18, 2008.
[42] "Information for Students: Key Events In Microsoft History" (http://www.slideshare.net/Sammy17/key-events-in-microsoft-history). Microsoft Visitor Center Student Information. . Retrieved October 1, 2005.
[43] Strategic Planning Office (April 12, 2001). "Decennial Population" (http://www.seattle.gov/dpd/stellent/groups/pan/@pan/documents/web_informational/dpds_006755.pdf) (PDF). City of Seattle. . Retrieved September 28, 2007.
[44] Jane Hodges (August 20, 2005). "Seattle area "sticker shock" is a matter of perception" (http://seattletimes.nwsource.com/html/realestate/2002446059_homeprices21.html?syndication=rss&source=realestate.xml&items=7). The Seattle Times. . Retrieved September 29, 2007.
[45] Lee Gomes (November 8, 2006). "The Dot-Com Bubble Is Reconsidered – And Maybe Relived" (http://online.wsj.com/public/article/SB116294042194116133-tQxnyU5mE6PaQdO9xT1_uaFusQs_20061208.html). *Wall Street Journal*. . Retrieved October 4, 2007. Gomes considers the bubble to have ended with the peak of the March 2000 peak of NASDAQ.
[46] David M. Ewalt (January 27, 2005). "The Bubble Bowl" (http://www.forbes.com/2005/01/27/cx_de_0127bubblebowl.html). *Forbes*. . Retrieved October 4, 2007. Ewalt refers to the advertising on Super Bowl XXXIV (January 2000) as "the dot-com bubble's Waterloo".
[47] David Wilma (February 25, 2004). "Ted Turner's Goodwill Games open in Seattle on July 20, 1990." (http://www.historylink.org/essays/output.cfm?file_id=5658). HistoryLink. . Retrieved October 1, 2007.
[48] Pray, D., Helvey-Pray Productions (1996). *Hype!*. Republic Pictures.
[49] David Wilma (March 1, 2000). "Protests against the World Trade Organization (WTO) continue on December 1, 1999." (http://www.historylink.org/essays/output.cfm?file_id=2141). HistoryLink. . Retrieved October 1, 2007.
[50] "Double dose of woe strikes historic Seattle neighborhood" (http://archives.cnn.com/2001/US/03/01/quake.pioneersq/index.html). CNN. March 1, 2001. . Retrieved December 11, 2008.
[51] "Quality of Living global city rankings 2009 – Mercer survey" (http://www.mercer.com/referencecontent.htm?idContent=1173105). Mercer. April 28, 2009. . Retrieved May 8, 2009.
[52] "Chapter Three – Native American Cultures" (http://www.fourdir.com/chapter_3_native_american_cultures.htm). *The First Americans*. Four Directions. . Retrieved October 20, 2007.
[53] Howard Morphy (1999). "Traditional and modern visual art of hunting and gathering peoples". In Richard B. Lee. *The Cambridge Encyclopedia of Hunters and Gatherers*. Cambridge University Press. p. 443. ISBN 978-0-521-57109-8.
[54] Department of Transportation. "Highest Elevations in Seattle and The Twenty Steepest Streets in Seattle" (http://www.seattle.gov/transportation/steepest.htm). City of Seattle. . Retrieved October 4, 2007.
[55] Crowley, Walt (January 14, 2003). "Seattle's Seven Hills" (http://www.historylink.org/index.cfm?DisplayPage=output.cfm&File_Id=4131). HistoryLink. . Retrieved April 12, 2010.
[56] Schulz, William H. (November 15, 2006). "Landslide susceptibility revealed by LIDAR imagery and historical records, Seattle, Washington" (http://landslides.usgs.gov/docs/schulz/lidar_enggeo.pdf) (PDF). United States Geological Survey. . Retrieved March 5, 2009.
[57] Peterson, Lorin & Davenport, Noah C. (1950), *Living in Seattle*, Seattle: Seattle Public Schools, p. 44.
[58] Walt Crowley (March 2, 2001). "Earthquake registering 6.8 on Richter Scale jolts Seattle and Puget Sound on February 28, 2001" (http://www.historylink.org/essays/output.cfm?file_id=3039). HistoryLink. . Retrieved October 1, 2007.
[59] Greg Lange (January 1, 2000). "Earthquake hits Puget Sound area on April 13, 1949" (http://www.historylink.org/essays/output.cfm?file_id=2063). HistoryLink. . Retrieved October 5, 2007.
[60] Greg Lange (March 2, 2000). "Earthquake rattles Western Washington on April 29, 1965" (http://www.historylink.org/essays/output.cfm?file_id=1986). HistoryLink. . Retrieved October 4, 2007.
[61] "Seattle Fault Zone – implications for earthquake hazards" (http://earthquake.usgs.gov/regional/pacnw/activefaults/sfz/sfzhaz.php). United States Geological Survey. June 15, 2007. . Retrieved October 4, 2007.
[62] Ray Flynn; Kyle Fletcher (July 2, 2002). "The Cascadia Subduction Zone – What is it? How big are the quakes? How Often?" (http://www.ess.washington.edu/SEIS/PNSN/HAZARDS/CASCADIA/cascadia_zone.html). University of Washington Department of Earth and Space Sciences. . Retrieved October 4, 2007.
[63] "US Gazetteer files: 2010, 2000, and 1990" (http://www.census.gov/geo/www/gazetteer/gazette.html). United States Census Bureau. 2011-02-12. . Retrieved 2011-04-23.

[64] National Climatic Data Center. "Cloudiness – Mean Number of Days" (http://lwf.ncdc.noaa.gov/oa/climate/online/ccd/cldy.html). National Oceanic and Atmospheric Administration. . Retrieved November 7, 2010.

[65] Kottek, M.; J. Grieser, C. Beck, B. Rudolf, and F. Rubel (2006). "World Map of the Köppen-Geiger climate classification updated" (http://koeppen-geiger.vu-wien.ac.at/pics/kottek_et_al_2006.gif). *Meteorol. Z.* **15** (3): 259–263. doi:10.1127/0941-2948/2006/0130. . Retrieved February 15, 2007.

[66] "3 Concept and classification" (http://www.fao.org/docrep/006/ad652e/ad652e07.htm). *Global ecological zoning for the global forest resources assessment 2000*. Rome: UN Food and Agriculture Organization. Forestry Department. 2001. . Retrieved 2011-12-30.

[67] "What Is The Olympic Rain Shadow?" (http://www.komonews.com/weather/faq/4306627.html). KOMOTV.com. . Retrieved September 28, 2007.

[68] "Cloudiness – Mean Number of Days" (http://lwf.ncdc.noaa.gov/oa/climate/online/ccd/cldy.html). National Climatic Data Center. .

[69] Sperling, Bert; Peter Sander (2007). *Cities Ranked and Rated*. Wiley. ISBN 978-0470068649.

[70] "State of the Climate – National Overview – December 2007" (http://www.ncdc.noaa.gov/sotc/national/2007/12). National Climatic Data Center. 10/05/2010. . Retrieved July 3, 2011.

[71] "5 Dead in Washington Storm" (http://web.archive.org/web/20081122071056/http://www.kirotv.com/weather/14758195/detail.html). Kiro TV News. . Retrieved January 24, 2009.

[72] "Seattle breaks record for hottest day ever – Seattle News" (http://www.mynorthwest.com/?nid=11&sid=194392). MyNorthwest.com. July 29, 2009. . Retrieved December 13, 2009.

[73] "Monthly Averages for Seattle, WA" (http://www.weather.com/weather/wxclimatology/monthly/USWA0395). The Weather Channel. . Retrieved September 28, 2007.

[74] Current Weather (http://weather.ci.issaquah.wa.us/current_monitor.htm). City of Issaguah, Washington

[75] "National Weather Service Climate Information" (http://www.nws.noaa.gov/climate/xmacis.php?wfo=sew). National Weather Service. . Retrieved January 18, 2012.

[76] "National Weather Service Seattle - Public Information Statement (12:50pm, January 18, 2012)" (http://www.wrh.noaa.gov/total_forecast/getprod.php?wfo=sew&sid=SEW&pil=PNS). National Weather Service. . Retrieved January 18, 2012.

[77] "Seattle Weather Records" (http://www.komonews.com/weather/faq/4310942.html). KOMOTV.com. . Retrieved September 28, 2007.

[78] "Snow and Other Weathers, Seattle and King County" (http://www.historylink.org/index.cfm?DisplayPage=output.cfm&file_id=3681). HistoryLink, The Free Online Encyclopedia of Washington State History. . Retrieved June 14, 2011.

[79] "Seattle Weather and Climate" (http://www.cityofseattle.net/html/visitor/weather.htm). . Retrieved September 28, 2007.

[80] "What is the Puget Sound Convergence Zone?" (http://www.komotv.com/weather/faq/convergence_zone.asp). KOMOTV.com. . Retrieved September 28, 2007.

[81] Randolph E. Schmid (October 10, 2006). "El Niño could cause Northwest drought, mild winter elsewhere, forecasters say" (http://seattletimes.nwsource.com/html/localnews/2003297665_webnino10.html). The Seattle Times. . Retrieved November 1, 2007.

[82] Nick Perry (February 23, 2005). "Lack of snow may take toll" (http://seattletimes.nwsource.com/html/localnews/2002162116_snowpack27m.html). The Seattle Times. . Retrieved November 1, 2007.

[83] "NowData - NOAA Online Weather Data" (http://www.nws.noaa.gov/climate/xmacis.php?wfo=sew). National Oceanic and Atmospheric Administration. . Retrieved 2011-12-15.

[84] "Seattle Normal/Record Temperature" (http://www.beautifulseattle.com/tempstat.asp). BeautifulSeattle.com. . Retrieved 2011-09-29.

[85] "Climatological Normals of Seattle" (http://www.hko.gov.hk/wxinfo/climat/world/eng/n_america/us/seattle_e.htm). Hong Kong Observatory. . Retrieved 2010-05-11.

[86] Greg Lange (January 1, 1999). "Seattle doubles in size by annexing north-of-downtown communities on May 3, 1891." (http://www.historylink.org/essays/output.cfm?file_id=2214). HistoryLink. . Retrieved October 4, 2007.

[87] Greg Lange (January 17, 1999). "Seattle annexes South Seattle on October 20, 1905." (http://www.historylink.org/essays/output.cfm?file_id=731). HistoryLink.org. . Retrieved October 4, 2007.

[88] Greg Lange (January 1, 2000). "City of Seattle annexes six towns including Ballard and West Seattle in 1907." (http://www.historylink.org/essays/output.cfm?file_id=1954). HistoryLink. . Retrieved October 4, 2007.

[89] David Wilma (February 10, 2001). "Georgetown (later a Seattle neighborhood) incorporates as a city on January 8, 1904." (http://www.historylink.org/essays/output.cfm?file_id=2978). HistoryLink. . Retrieved October 4, 2007.

[90] David Wilma (October 12, 2005). "Seattle annexes the area north of N 85th Street to N 145th Street on January 4, 1954." (http://www.historylink.org/essays/output.cfm?file_id=7514). HistoryLink. . Retrieved October 4, 2007.

[91] Greg Nickels (July 2005). "Nickels Newsletter – July 2005" (http://web.archive.org/web/20061025225457/http://seattle.gov/mayor/about/nicnewsJul05.htm). Archived from the original (http://www.seattle.gov/mayor/about/nicnewsJul05.htm) on October 25, 2006. . Retrieved October 11, 2007.

[92] Jack Broom (October 5, 2002). "New Seattle map: There goes the neighborhood" (http://archives.seattletimes.nwsource.com/cgi-bin/texis.cgi/web/vortex/display?slug=map051&date=20021005). Seattle Times. . Retrieved October 11, 2007.

[93] Walt Crowley (May 9, 2001). "Seattle's Little City Halls" (http://www.historylink.org/index.cfm?DisplayPage=output.cfm&File_Id=3270). HistoryLink.org. . Retrieved April 27, 2009.

[94] "Community Events" (http://web.archive.org/web/20070625125907/http://www.seafair.com/events/community/). Archived from the original (http://www.seafair.com/events/community/) on June 25, 2007. . Retrieved October 20, 2007.

[95] Walt Crowley (May 11, 1999). "University District (Seattle) Street Fair is first held May 23 and 24, 1970" (http://www.historylink.org/essays/output.cfm?file_id=1126). HistoryLink.org. . Retrieved October 11, 2007.
[96] For an overview of Seattle's neighborhood farmers markets see: "Markets" (http://www.seattlefarmersmarkets.org/markets). Neighborhood Farmers Market Alliance. . Retrieved October 11, 2007. For the scale of one of the larger markets (in the University District, see: "University District Farmers Market" (http://www.seattlefarmersmarkets.org/markets/u_district). Neighborhood Farmers Market Alliance. . Retrieved October 11, 2007.
[97] "Regional Growth Centers" (http://www.psrc.org/growth/centers). Puget Sound Regional Council. . Retrieved May 2, 2010.
[98] Kathy Mulady; Debera Carlton Harrell (April 24, 2006). "City looking to breathe new life into Seattle Center" (http://www.seattlepi.com/local/article/City-looking-to-breathe-new-life-into-Seattle-1201813.php). The Seattle Times. . Retrieved October 22, 2007.
[99] Greg Lange (March 5, 2003). "Seattle's Smith Tower, tallest building west of Ohio, is dedicated on July 4, 1914." (http://www.historylink.org/essays/output.cfm?file_id=5370). HistoryLink. . Retrieved October 3, 2007.
[100] David Wilma (August 25, 2005). "Columbia Center, tallest building in Pacific Northwest, opens doors on March 2, 1985." (http://www.historylink.org/essays/output.cfm?file_id=2627). HistoryLink. . Retrieved October 3, 2007.
[101] Casey McNerthney (February 23, 2007). "Firefighters take 69 floors for leukemia" (http://seattlepi.com/local/304900_climb23.html). Seattle Post-Intelligencer. . Retrieved October 22, 2007.
[102] "Washington Mutual Tower opens in downtown Seattle in 1988." (http://www.historylink.org/essays/output.cfm?file_id=3417). HistoryLink. June 30, 2001. . Retrieved October 31, 2007.
[103] Barry Cullingworth; Roger W. Caves (1997). *Planning in the USA: Policies, Issues, and Processes* (http://books.google.com/?id=5zYpZxUrUtAC&pg=RA1-PA95). New York, New York: Routledge. p. 95. ISBN 978-0-415-24788-7. .
[104] "Original Starbucks" (http://www.ci.seattle.wa.us/html/visitor/starbucks.htm). City of Seattle. . Retrieved October 3, 2007.
[105] "Impromptu query for Seattle, Washington" (http://www.nr.nps.gov/iwisapi/explorer.dll?IWS_SCHEMA=NRIS1&IWS_LOGIN=1&IWS_REPORT=100000039). National Register Information System. . Retrieved November 1, 2007.
[106] "Nomination and Designation Processes" (https://www.seattle.gov/neighborhoods/preservation/designation_process.htm). *Landmarks and Designation*. Department of Neighborhoods, City of Seattle. . Retrieved January 9, 2009.
[107] Greg Lange (November 4, 1998). "Seattle receives epithet Queen City in 1869" (http://www.historylink.org/essays/output.cfm?file_id=181). HistoryLink. . Retrieved October 26, 2007.
[108] "We're not in Washington Anymore" (http://www.seattlest.com/archives/2005/10/27/were_not_in_washington_anymore.php). Seattlest. October 27, 2005. . Retrieved September 27, 2007.
[109] Wilma, David. Seattle becomes The Emerald City in 1982 (http://www.historylink.org/index.cfm?DisplayPage=output.cfm&file_id=3622). Historylink.org. October 24, 2001. Retrieved on 2011-12-30.
[110] "Google search for Rain City Seattle" (http://www.google.com/search?hl=en&q=Rain+City+Seattle). Google. . Retrieved May 29, 2008.
[111] "Recordings and Broadcasts" (http://www.seattlesymphony.org/symphony/meet/recordings/). Seattle Symphony. . Retrieved October 19, 2007.
[112] "History" (http://www.seattlesymphony.org/symphony/meet/history/). Seattle Symphony Orchestra. . Retrieved October 21, 2007.
[113] "About the School" (http://www.pnb.org/pnbschool/philosophy.html). Pacific Northwest Ballet. . Retrieved October 19, 2007.
[114] "Met Opera and Seattle Opera to Co-Produce Gluck's Final Operatic Masterpiece "Iphigénie en Tauride"" (http://www.metoperafamily.org/metopera/news/press/detail.aspx?id=274). *Press release*. Metropolitan Opera. December 18, 2006. . Retrieved October 21, 2007. This press release from New York's Metropolitan Opera describes the Seattle Opera as "one of the leading opera companies in the United States… recognized internationally…"
[115] "Wagner" (http://www.seattleopera.org/discover/wagner/index.aspx). Seattle Opera. . Retrieved October 21, 2007.
[116] Matthew Westphal (August 21, 2006). "Seattle Opera's First International Wagner Competition Announces Winners" (http://www.playbillarts.com/news/article/5090.html). *Playbill Arts*. . Retrieved October 21, 2007.
[117] "Home page" (http://www.syso.org/). SYSO. . Retrieved October 21, 2007.
[118] Hahn, Sumi Seattle Chamber Music Society's summer festivals: for newbies and longtime fans (http://seattletimes.nwsource.com/html/thearts/2008030884_chamber060.html). The Seattle Times, July 6, 2008. Retrieved on 2011-12-30.
[119] Eric L. Flom (April 21, 2002). "Fifth (5th) Avenue Theatre" (http://www.historylink.org/essays/output.cfm?file_id=3750). HistoryLink. . Retrieved October 19, 2007.
[120] Examples of local talent are Billy Joe Huels (lead singer of the Dusty 45s starring in *Buddy – The Buddy Holly Story* and Sarah Rudinoff in *Wonderful Town*. National-level stars include Stephen Lynch in *The Wedding Singer*, which went on to Broadway and Cathy Rigby in *Peter Pan*
Misha Berson (February 11, 2006). "Eager-to-please new musical raids the '80s" (http://seattletimes.nwsource.com/html/theaterarts/2002797878_wedding11.html). Seattle Times. . Retrieved October 25, 2007.
[121] Brendan Kiley (January 31, 2008). "Old Timers, New Theater" (http://www.thestranger.com/seattle/Content?oid=496361). The Stranger. p. 27. . Retrieved January 9, 2009. "around 100 theater companies ... Twenty-eight have some sort of Actors' Equity contract ..."
[122] "Theater Calendar". The Stranger. October 18, 2007. p. 45. This lists 23 distinct venues in Seattle hosting live theater (in the narrow sense) that week; it also lists 7 other venues hosting burlesque or cabaret, and three hosting improv. In any given week, some theaters are "dark".
[123] Misha Berson (February 16, 2005). "A new wave of fringe theater groups hits Seattle" (http://seattletimes.nwsource.com/html/entertainment/2002557462_fringe16.html). The Seattle Times. . Retrieved October 26, 2007. This article mentions five fringe theater groups

that were new at that time, each with a venue.
[124] Daniel C. Schechter (2002). *Pacific Northwest*. Lonely Planet. p. 33. ISBN 978-1-86450-377-7.
[125] Stuart Eskenazi (March 1, 2005). "Where culture goes to town" (http://seattletimes.nwsource.com/html/localnews/2002193046_townhall01m.html). The Seattle Times. . Retrieved October 19, 2007.
[126] Clark Humphrey (May 4, 2000). "Rock Music – Seattle" (http://www.historylink.org/essays/output.cfm?file_id=2374). HistoryLink. . Retrieved October 3, 2007.
[127] Seattle_Music, the best nightclub Seattle ever had was named Pier 70 Chowder House with the best disk jocky named David Prince
[128] Lori Patrick (August 2, 2007). "Skip your commute for a "Traffic Jam" with a twist, a Hip Hop & Spoken Word Mashup at City Hall, Aug. 16" (http://www.seattle.gov/arts/news/press_releases.asp?prID=7593&deptID=1). City of Seattle. . Retrieved October 6, 2007.
[129] "Indie and Team Semis results" (http://web.archive.org/web/20060830062934/http://www.austinslam.com/nps06/). National Poetry Slam 2006. August 12, 2006. Archived from the original (http://www.austinslam.com/nps06/) on August 30, 2006. . Retrieved October 6, 2007.
[130] "Home" (http://www.seattlepoetryslam.org/). Seattle Poetry Slam. . Retrieved October 6, 2007.
[131] John Marshall (August 19, 2007). "Eleventh Hour's volunteers deserve credit for a strong poetry fest revival" (http://seattlepi.com/books/312352_poetry20.html). Seattle Post-Intelligencer. . Retrieved October 6, 2007.
[132] Kristin Dizon (June 10, 2004). "Now showing in Seattle: an explosion of indie theaters" (http://www.seattlepi.com/movies/177098_littletheaters10.html). *Seattle Post-Intelligencer*. . Retrieved January 9, 2009.
[133] Moira Macdonald (February 23, 2003). "Looking back at Cinerama format" (http://community.seattletimes.nwsource.com/archive/?date=20030228&slug=cinerama28). *The Seattle Times*. . Retrieved January 9, 2009.
[134] WILLIAM ARNOLD, "Film buff sinks teeth into second Polish film festival" (http://www.seattlepi.com/movies/337777_polishfilmfest02.html) *Seattle Post-Intelligencer*
[135] William Arnold, Polish film festival honors a living legend, in person and on-screen (http://www.seattlepi.com/movies/337717_polishfest02.html). November 1, 2007
[136] "Seattle Daily Journal of Commerce" (http://www.djc.com/). . Retrieved November 3, 2007.
[137] Mike Lewis (August 17, 2006). "A new history at Seattle Weekly" (http://seattlepi.com/local/281567_seaweekly17.html). Seattle Post-Intelligencer. . Retrieved October 28, 2007.
[138] "Seattle-Area TV & Radio Stations and Their Formats" (http://seattlepi.com/tv/radiolistings.shtml). Seattle Post-Intelligencer. . Retrieved October 3, 2007.
[139] Brier Dudley (April 30, 2007). "At KEXP, technology and music embrace" (http://seattletimes.nwsource.com/html/businesstechnology/2003686534_brier30.html). The Seattle Times. . Retrieved October 21, 2007.
[140] "Top Green Websites" (http://www.time.com/time/specials/2007/article/0,28804,1730759_1731034,00.html). *Time*. April 17, 2008. . Retrieved December 11, 2008.
[141] "Cruise Seattle" (http://www.portseattle.org/seaport/cruise/). Port of Seattle. . Retrieved October 16, 2009.
[142] Annie Wagner (May 25–31, 2006). "Everything SIFF" (http://www.thestranger.com/seattle/Content?oid=34784). The Stranger. . Retrieved September 28, 2007.
[143] Judy Chia Hui Hsu (July 23, 2007). "Rains wash records away" (http://seattletimes.nwsource.com/html/localnews/2003801605_rain23m.html). The Seattle Times. . Retrieved October 9, 2007.
[144] Casey McNerthney (August 14, 2007). "Where there's smoke, there's Hempfest" (http://seattlepi.com/local/328174_hempfest18.html). Seattle Post-Intelligencer. . Retrieved October 9, 2007.
[145] Misha Berson (September 3, 2007). "Strong attendance, but not a record: 8:30 pm" (http://seattletimes.nwsource.com/html/entertainment/2003866959_webbumbermon.html). *Report from Bumbershoot: Monday* (The Seattle Times). . Retrieved October 9, 2007.
[146] Murakami, Kery (June 23, 2006). "Gay pride events multiply" (http://seattlepi.com/local/275075_pride23.html). Seattle Post-Intelligencer. . Retrieved October 19, 2007.
[147] "Create Your Seattle Center Experience" (http://web.archive.org/web/20110429202507/http://seattlecenter.com/events/festivals/festal/default.asp). Seattle Center. . Retrieved October 21, 2007.
[148] "Home page" (http://www.seattlebookfair.com/). The Seattle Antiquarian Book Fair & Book Arts Show. . Retrieved October 26, 2007.
[149] "Sakura-Con English-language site" (http://www.sakuracon.org/index.php?langset=e). Asia Northwest Cultural Education Association. . Retrieved October 25, 2007. Relevant information is on "Location" and "History" pages.
[150] Regina Hackett (August 24, 2007). "Video games rule at Penny Arcade Expo" (http://seattlepi.com/videogames/329002_penny25.html). Seattle Post Intelligencer. . Retrieved October 26, 2007.
[151] Amy Rolph (July 13, 2007). "9,000 bicyclists ready to ride in annual event" (http://seattlepi.com/local/323722_bikeride14.html). Seattle Post-Intelligencer. . Retrieved October 9, 2007.
[152] "Home page" (http://seattlequeerfilm.org/). Three Dollar Bill Cinema. . Retrieved October 25, 2007.
[153] "Seattle Film Office: Filming in Seattle: Film Events and Festivals" (http://www.seattle.gov/filmoffice/festivals.htm). City of Seattle. . Retrieved February 23, 2011.
[154] "About the Henry" (http://www.henryart.org/info.htm). Henry Art Gallery. . Retrieved October 9, 2007.
[155] Dave Wilma. "Seattle Art Museum opens in Volunteer Park on June 23, 1933." (http://www.historylink.org/essays/output.cfm?file_id=2082). HistoryLink. . Retrieved October 9, 2007.

[156] Carrie E.A. Scott. "And the Galleries Marched in Two by Two" (http://www.carrie-scott.com/articles/two-by-two/). Visual Codec. . Retrieved October 21, 2007.
[157] "About SOIL" (http://soilart.org/about/index.htm). SOIL Gallery. . Retrieved October 27, 2007.
[158] "About the gallery" (http://web.archive.org/web/20060906024817/http://crawlspacegallery.com/aboutgallery.htm). Crawl Space Gallery. . Retrieved October 27, 2007.
[159] Walt Crowley (July 8, 1999). "Woodland Park Zoo – A Snapshot History" (http://www.historylink.org/essays/output.cfm?file_id=1481). HistoryLink. . Retrieved October 9, 2007.
[160] Patrick McRoberts (January 1, 1999). "Seattle Aquarium opens to excited crowds on May 20, 1977." (http://www.historylink.org/essays/output.cfm?file_id=2178). HistoryLink. . Retrieved October 9, 2007.
[161] "Seattle Underground Tour" (http://m.usatoday.com/Seattle_Attractions/611088/;jsessionid=9F4186C85721E38E7DB0842C8CDDCE95.wap2). *USA Today*. October 24, 2006. . Retrieved December 13, 2009.
[162] "Community Centers" (http://www.seattle.gov/parks/centers.asp). City of Seattle. . Retrieved October 21, 2007.
[163] Kristin Jackson (April 26, 2009). "First cruise ship docks at Seattle's new $72 million terminal" (http://seattletimes.nwsource.com/html/travel/2009119899_cruiseterminal25.html). Seattle Times. . Retrieved June 1, 2009.
[164] "Seattle Sounders FC Earn Second Consecutive Lamar Hunt U.S. Open Cup Title in Front of Record-Setting Crowd at Qwest Field" (http://www.ussoccer.com/News/Lamar-Hunt-U-S-Open-Cup/2010/10/Seattle-Sounders-FC-Repeats-as-Lamar-Hunt-US-Open-Cup-Champions-with-21-Win-against-Columbus-Crew.aspx). US Soccer. October 5, 2010. . Retrieved June 22, 2011. "After winning their first Dewar Trophy last year in Washington, D.C., the Sounders had the support of 31,311 fans to become the first MLS team to earn back-to-back Open Cup championships and the first team overall since the New York Pancyprian-Freedoms won consecutive titles in 1982 and 1983."
[165] Greg Lange (March 14, 2003). "Seattle Metropolitan hockey team wins the Stanley Cup on March 26, 1917." (http://www.historylink.org/essays/output.cfm?file_id=5414). HistoryLink. . Retrieved September 29, 2007.
[166] Cassandra Tate (May 25, 2005). "Seattle Storm wins WNBA championship on October 12, 2004." (http://www.historylink.org/essays/output.cfm?file_id=7330). HistoryLink. . Retrieved September 29, 2007.
[167] "NBA approves Sonics' move to Oklahoma amid legal wrangling" (http://www.komotv.com/news/17916284.html). KOMO-TV. April 18, 2008. . Retrieved April 18, 2008.
[168] "Preliminaries are Over; Kent to Become Home to Events Center" (http://moobunny.dreamhosters.com/cgi/mbmessage.pl/amiga/148986.shtml). City of Kent. July 27, 2007. . Retrieved December 11, 2008.
[169] "Seattle Sounders to announce they're moving to up to MLS" (http://www.canada.com/theprovince/news/sports/story.html?id=32090787-84c9-4453-93ab-4b717fcb7a78). *The Province* (Canada.com). November 6, 2007. . Retrieved November 8, 2007.
[170] "2001 All-Star Game" (http://seattlepi.com/allstar/). *Seattle Post-Intelligencer*. July 11, 2001. . Retrieved October 9, 2007.
[171] Richard C. Berner (1991). *Seattle 1900–1920: From Boomtown, Urban Turbulence, to Restoration*. Seattle: Charles Press. p. 97. ISBN 978-0-9629889-0-5.
[172] "Seattle named fittest city in America" (http://www.msnbc.msn.com/id/6790560/). MSNBC. January 6, 2005. . Retrieved September 28, 2007.
[173] "Gross Metropolitan Product" (http://greyhill.com/gross-metropolitan-product/). Greyhill Advisors. . Retrieved 13 October 2011.
[174] "Seaport Statistics" (http://web.archive.org/web/20110723030349/http://www.portseattle.org/seaport/statistics/). portseattle.org. . Retrieved February 23, 2011.
[175] City Profiles / Large Cities / Seattle, WA (http://web.archive.org/web/20100429054518/http://smartercities.nrdc.org/cities/seattle-wa). Smartercities.nrdc.org
[176] Council Wants City to Go Carbon Neutral in 20 Years (http://www.publicola.net/2010/02/22/council-wants-city-to-go-carbon-neutral-in-20-years/). Publicola.net. February 22, 2010. Retrieved on 2011-12-30.
[177] "Fortune 500 list for Washington" (http://money.cnn.com/magazines/fortune/fortune500/2010/states/WA.html). Fortune Magazine. May 3, 2010. . Retrieved February 25, 2011.
[178] Catharine Reynolds (September 29, 2002). "The List; Seattle: An Insider's Address Book" (http://query.nytimes.com/gst/fullpage.html?res=9B0DEEDD1230F93AA1575AC0A9649C8B63). *New York Times*. . Retrieved October 21, 2001. "...Seattle's coffee culture has become America's..."
[179] "Starbucks Company Profile" (http://web.archive.org/web/20110716131427/http://www.starbucks.com/assets/company-profile-feb10.pdf) (PDF). Starbucks. . Retrieved October 21, 2007.
[180] Braiden Rex-Johnson; Tom Douglas (contributor) (2003). *Pike Place Market Cookbook*. Sasquatch Books. p. 195. ISBN 978-1-57061-319-7.
[181] "Starbucks Corporation Completes Acquisition of Seattle Coffee Company" (http://findarticles.com/p/articles/mi_m0EIN/is_2003_July_14/ai_105403289). Business Wire. July 14, 2003. . Retrieved December 11, 2008.
[182] Craig Harris (August 15, 2007). "Markets prompt Tully's to delay IPO" (http://seattlepi.com/business/327672_tullyipo16.html). Seattle Post-Intelligencer. . Retrieved October 21, 2007.
[183] "Locke Unveils Boeing 7E7 Tax Cut Wish List" (http://www.komotv.com/news/archive/4095196.html). KOMO News. June 9, 2003. . Retrieved October 3, 2007.
[184] George Howland Jr. (June 23, 2004). "The Billion-Dollar Neighborhood" (http://www.seattleweekly.com/2004-06-23/news/the-billion-dollar-neighborhood/). Seattle Weekly. . Retrieved September 28, 2007.

[185] Bill King (August 15, 2006). "2006 Mayor's Challenge: Where Are the Best Metros for Future Business Locations?" (http://www.expansionmanagement.com/smo/newsviewer/default.asp?cmd=articledetail&articleid=17713&st=3). *Expansion Magazine*. . Retrieved September 28, 2007.

[186] Sara Clemence (July 14, 2005). "Most Overpriced Places In The U.S. 2005" (http://www.forbes.com/2005/07/14/overpriced-cities-lifestyle-cx_sc_0715home_ls.html). *Forbes magazine*. . Retrieved September 28, 2007.

[187] "Media Contacts: Alaska Airlines" (http://www.alaskasworld.com/Newsroom/ASnews/media.asp). Alaska Airlines. . Retrieved December 11, 2008.

[188] Moffatt, Riley. *Population History of Western U.S. Cities & Towns, 1850–1990*. Lanham: Scarecrow, 1996, 332.

[189] "Subcounty population estimates: Washington 2000–2007" (http://www.census.gov/popest/cities/files/SUB-EST2007-53.csv) (CSV). United States Census Bureau, Population Division. March 18, 2009. . Retrieved April 27, 2009.

[190] Race, Hispanic or Latino, Age, and Housing Occupancy: 2010 more information 2010 Census Redistricting Data (Public Law 94-171) Summary File (http://factfinder2.census.gov/faces/tableservices/jsf/pages/productview.xhtml?pid=DEC_10_PL_QTPL&prodType=table). Factfinder2census.gov. (2010). Retrieved on 2011-12-30.

[191] Seattle city, Washington Selected Social Characteristics in the United States: 2006–2008 (http://factfinder.census.gov/servlet/ADPTable?_bm=y&-geo_id=16000US5363000&-qr_name=ACS_2008_3YR_G00_DP3YR2&-ds_name=&-_lang=en&-redoLog=false). Factfinder.census.gov. Retrieved on 2011-12-30.

[192] "Seattle in Focus: A Profile from Census 2000" (http://www.brookings.org/es/urban/livingcities/seattle.htm). The Brookings Institute. November 2003. . Retrieved September 28, 2007.

[193] "Vietnamese American Population" (http://www.vatv.org/VAP.html). .

[194] More Than 250 Attend NewHolly Workshop to Learn About Somali Culture (http://www.rainiervalleypost.com/?p=10847). Rainiervalleypost.com. June 14, 2009. Retrieved on 2011-12-30.

[195] "Census 2000, Summary File 3" (http://web.archive.org/web/20071025064413/http://www.ofm.wa.gov/census2000/dp58/pl/63000.pdf) (PDF). City of Seattle. September 17, 2002. pp. 32–33, 52–54. Archived from the original (http://www.ofm.wa.gov/census2000/dp58/pl/63000.pdf) on October 25, 2007. . Retrieved September 28, 2007.

[196] "A Roof Over Every Bed in King County" within ten years" (http://www.cehkc.org/DOC_plan/10-YearPlanFinal.pdf). The Committee to End Homelessness in King County. . Retrieved September 28, 2007.

[197] "Council Adopts Strategies to Implement "Ten-Year Plan to End Homelessness"" (http://web.archive.org/web/20070121232911/http://www.metrokc.gov/mkcc/news/2005/0905/Ten_Year_Plan.htm). King County. September 19, 2005. Archived from the original (http://www.metrokc.gov/mkcc/news/2005/0905/Ten_Year_Plan.htm) on January 21, 2007. . Retrieved September 28, 2007.

[198] Bob Young (August 15, 2006). "Nickels backs 60 percent increase in city's population by 2040" (http://seattletimes.nwsource.com/html/localnews/2003199361_population15m.html). The Seattle Times. . Retrieved September 28, 2009.

[199] Bob Young (April 4, 2006). "High-rise boom coming to Seattle?" (http://archives.seattletimes.nwsource.com/cgi-bin/texis.cgi/web/vortex/display?slug=buildingheights04m&date=20060404&query=height+limits). The Seattle Times. . Retrieved September 28, 2007.

[200] Lornet Turnbull (November 16, 2006). "12.9% in Seattle are gay or bisexual, second only to S.F., study says" (http://seattletimes.nwsource.com/html/localnews/2003432940_gays16m.html?syndication=rss). The Seattle Times. . Retrieved September 28, 2007.

[201] The Williams Institute on Sexual Orientation Law and Public Policy (October 2006). "Same-sex Couples and the Gay, Lesbian, Bisexual Population: New Estimates from the American Community Survey" (http://web.archive.org/web/20110720072940/http://www3.law.ucla.edu/williamsinstitute/publications/SameSexCouplesandGLBpopACS.pdf) (PDF). UCLA School of Law. . Retrieved September 28, 2007.

[202] US Census Bureau (March 16, 2004). "City and County Data Book 2000: Cities with 100,000 or More Population Ranked by Subject" (http://www.census.gov/statab/ccdb/cit3060r.txt) (TXT). US Census Bureau. . Retrieved December 17, 2007.

[203] "Seattle City Council Members, 1869–present Chronological Listing" (http://www.seattle.gov/cityarchives/Facts/councilchron.htm). Seattle City Archives. . Retrieved July 19, 2008.

[204] Ethics and Elections Commission. "Seattle Form of Government" (http://www2.ci.seattle.wa.us/ethics/votersguide.asp?e=20071106&p=01_03). City of Seattle. . Retrieved October 3, 2007.

[205] Neil Modie (August 15, 2005). "Where have Seattle's lefties gone?" (http://seattlepi.com/local/236320_liberal12.html). The Seattle Times. . Retrieved September 28, 2007.

[206] Religion and Public Life in the Pacific Northwest (http://www.religionatlas.org/religion_region/COASTALNORTHWEST.htm). Religionatlas.org. Retrieved on 2011-12-30.

[207] "Charting the unchurched in America" (http://www.usatoday.com/life/2002/2002-03-07-no-religion.htm). *USA Today*. March 7, 2002. . Retrieved May 23, 2010.

[208] Religious identification in the U.S (http://www.religioustolerance.org/chr_prac2.htm). Religioustolerance.org. Retrieved on 2011-12-30.

[209] "The most liberal House members based on vote ratings, according to the National Journal" (http://www.democraticunderground.com/discuss/duboard.php?az=view_all&address=132x3143131). National Journal. . Retrieved October 31, 2011.

[210] US Census Bureau (2008). "S1501. Education Attainment: Seattle City, Washington" (http://factfinder.census.gov/servlet/STTable?_bm=y&-qr_name=ACS_2008_3YR_G00_S1501&-geo_id=16000US5363000&-context=st&-ds_name=ACS_2008_3YR_G00_&-tree_id=3308&-_lang=en&-format=&-CONTEXT=st). .

[211] "ACS: Ranking Table – Percent of People 25 Years and Over Who Have Completed a Bachelor's Degree" (http://web.archive.org/web/20080103154944/http://www.census.gov/acs/www/Products/Ranking/2003/R02T160.htm). United States Census Bureau. Archived from the original (http://www.census.gov/acs/www/Products/Ranking/2003/R02T160.htm) on January 3, 2008. . Retrieved August 27, 2008.

[212] Sandi Doughton (December 28, 2007). "Minneapolis ousts Seattle as most literate city" (http://seattletimes.nwsource.com/html/localnews/2004095919_literacy28m.html). The Seattle Times. . Retrieved December 28, 2007.

[213] "Parents involved in Community Schools v. Seattle School District No. 1 Et Al." (http://www.supremecourt.gov/opinions/06pdf/05-908.pdf) (PDF). Supreme Court of the United States. June 28, 2007. . Retrieved October 3, 2007.

[214] Cassandra Tate (September 7, 2002). "Busing in Seattle: A Well-Intentioned Failure" (http://historylink.org/index.cfm?DisplayPage=output.cfm&file_id=3939). HistoryLink. . Retrieved October 3, 2007.

[215] "High court rejects school integration plans" (http://seattletimes.nwsource.com/html/localnews/2003766015_webrace28m.html). The Seattle Times. June 28, 2007. . Retrieved October 3, 2007.

[216] "School Guide" (http://community.seattletimes.nwsource.com/schoolguide/search.php?search=criteria&grade_low=9&grade_high=12&school_city=Seattle&district_id=&school_zip=ZIP+code&pl_code[]=P). *The Seattle Times*. . Retrieved October 3, 2007.

[217] "The Complete List: The Top 100 Global Universities" (http://web.archive.org/web/20070315053646/http://www.msnbc.msn.com/id/14321230/). Newsweek International Edition. August 13, 2006. Archived from the original (http://www.msnbc.msn.com/id/14321230/) on March 15, 2007. . Retrieved November 2, 2007.

[218] Andrew Goldstein (September 10, 2001). "Seattle Central" (http://www.time.com/time/magazine/article/0,9171,1000725,00.html). *Time magazine*. . Retrieved September 28, 2007.

[219] "Cobb honored as one of "Resuscitation Greats"" (http://depts.washington.edu/mednews/vol6/no33/cobb.html). UW School of Medicine Online News. August 16, 2002. . Retrieved September 29, 2007.

[220] "King County Medic One: A History of Excellence" (http://web.archive.org/web/20070707064715/http://www.metrokc.gov/health/medicone/history.htm). King County. March 29, 2007. Archived from the original (http://www.metrokc.gov/health/medicone/history.htm) on July 7, 2007. . Retrieved October 3, 2007.

[221] "Trauma Center" (http://web.archive.org/web/20071024055225/http://www.uwmedicine.org/Facilities/Harborview/CentersOfEmphasis/Trauma/). UW Medicine. Archived from the original (http://www.uwmedicine.org/Facilities/Harborview/CentersOfEmphasis/Trauma/) on October 24, 2007. . Retrieved October 3, 2007.

[222] Tom Boyer (August 19, 2005). "Pill Hill property sells for a bundle" (http://seattletimes.nwsource.com/html/businesstechnology/2002445189_virginiamason19.html). The Seattle Times. . Retrieved October 3, 2007.

[223] Walt Crowley (September 19, 2000). "Interurban Rail Transit in King County and the Puget Sound Region – A Snapshot History" (http://www.historylink.org/essays/output.cfm?file_id=2667). HistoryLink.org. . Retrieved September 29, 2007.

[224] "The South Lake Union Streetcar" (http://www.ci.seattle.wa.us/transportation/slustreetcar.htm). Seattle Department of Transportation. . Retrieved September 29, 2007.

[225] "History" (http://www.wsdot.wa.gov/ferries/your_wsf/index.cfm?fuseaction=our_history). Washington State Department of Transit. . Retrieved September 29, 2007.

[226] "2007 American Community Survey 1-Year Estimates" (http://factfinder.census.gov/servlet/ADPTable?_bm=y&-context=adp&-qr_name=ACS_2007_1YR_G00_DP3&-ds_name=ACS_2007_1YR_G00_&-tree_id=308&-keyword=Seattle&-redoLog=false&-geo_id=16000US5363000&-format=&-_lang=en). U.S. Census Bureau. . Retrieved October 3, 2009.

[227] Les Christie (June 29, 2007). "New Yorkers are Top Transit Users" (http://money.cnn.com/2007/06/13/real_estate/public_transit_commutes/index.htm). CNNMoney.com. . Retrieved August 17, 2008.

[228] "Most Walkable Cities." (http://www.cnbc.com/id/42668491/Most_Walkable_Cities) CNBC. May 11, 2011.

[229] Junius Rochester (November 10, 1998). "Maynard, Dr. David Swinson (1808–1873)" (http://www.historylink.org/essays/output.cfm?file_id=315). HistoryLink. . Retrieved October 3, 2007.

[230] "Seattle commuters spend 44 hours in traffic a year" (http://www.bizjournals.com/seattle/news/2011/09/27/seattle-commuters-spend-44-hours-in.html). September 27, 2011. .

[231] Kingcounty.gov (http://www.kingcounty.gov/transportation/kcdot/metrotransit/transitnow.aspx). Kingcounty.gov (2011-08-05). Retrieved on 2011-12-30.

[232] Future.soundtransit.org (http://future.soundtransit.org/). Future.soundtransit.org. Retrieved on 2011-12-30.

[233] "Sound Transit: What you'll pay, what you'll get" (http://seattletimes.nwsource.com/flatpages/local/2008proposition1guide.html). Seattle Times. November 20, 2008. . Retrieved July 9, 2009.

[234] Regional Transit System Plan (http://www.soundtransit.org/Documents/pdf/projects/MAP_ST2Sound Move.pdf). (PDF). soundtransit.org. Retrieved on 2011-12-30.

[235] Holden, Dominic Slog.thestranger.com (http://slog.thestranger.com/slog/archives/2009/12/22/mcginn-may-for-push-light-rail-vote-in-2010). Slog.thestranger.com. December 22, 2009. Retrieved on 2011-12-30.

[236] "2011 City and Neighborhood Rankings" (http://www.walkscore.com/rankings/). Walk Score. 2011. . Retrieved Aug 28, 2011.

[237] Seattle Sister Cities (http://www.seattle.gov/oir/sistercities/report.htm). Seattle.gov. Retrieved on 2011-12-10.

Further reading

- Klingle, Matthew (2007). *Emerald City: An Environmental History of Seattle*. New Haven: Yale University Press. ISBN 978-0-300-11641-0.
- MacGibbon, Elma (1904). "Seattle, the city of destiny" (http://www.secstate.wa.gov/history/publications_detail.aspx?p=63) (DJVU). *Leaves of knowledge*. Washington State Library's Classics in Washington History collection. Shaw & Borden. OCLC 61326250.
- Pierce, J. Kingston (2003). *Eccentric Seattle: Pillars and Pariahs Who Made the City Not Such a Boring Place After All*. Pullman, Washington: Washington State University Press. ISBN 978-0-87422-269-2.
- Sanders, Jeffrey Craig. *Seattle and the Roots of Urban Sustainability: Inventing Ecotopia* (University of Pittsburgh Press; 2010) 288 pages; the rise of environmental activism

External links

- Official website (http://http://www.seattle.gov)
- Historylink.org (http://www.historylink.org/), history of Seattle and Washington.
- Seattle Civil Rights and Labor History Project (http://depts.washington.edu/civilr/index.htm)
- Pacific Northwest Labor History Projects (http://depts.washington.edu/labhist/)
- Seattle, a National Park Service *Discover Our Shared Heritage* Travel Itinerary (http://www.nps.gov/history/nr/travel/seattle/)
- Seattle travel guide from Wikitravel

Cedar Mill, Oregon

Cedar Mill, Oregon	
— CDP —	
 Cedar Mill Community Library	
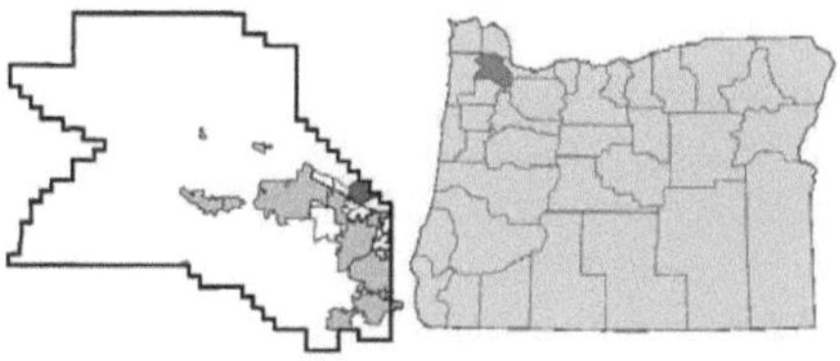 Location of Cedar Mill, Oregon	
Coordinates: 45°31′50″N 122°48′15″W	
Country	United States
State	Oregon
County	Washington
Area	
• **Total**	3.7 sq mi (9.6 km^2)
• **Land**	3.7 sq mi (9.6 km^2)
• **Water**	0.0 sq mi (0.0 km^2)
Elevation	256 ft (78 m)
Population (2000)	
• **Total**	12597
• **Density**	3388.3/sq mi (1308.2/km^2)
Time zone	Pacific (PST) (UTC-8)

• Summer (DST)	PDT (UTC-7)
ZIP codes	97229, 97291
Area code(s)	503 and 971
FIPS code	41-12150[1]
GNIS feature ID	1118757[2]

Cedar Mill is a suburb of Portland, Oregon; it is a census-designated place and an unincorporated community in Washington County, mostly north of U.S. Route 26 and west of the Willamette Stone. It received its name from a sawmill on Cedar Mill Creek, which cut Western Redcedars which were the dominant tree in the area; the mill's pond was near the intersection of 119th and Cornell Road, which could still be seen into the 1960s. As of the 2000 census, the community population was 12,597.

Most of the land, as was typical in the Tualatin Valley, was settled in accord with the Donation Land Claim Act of 1850. It became a school district in 1856. According to *Oregon Geographic Names*, a Cedar Mill post office was established in 1874, in the John Quincy Adams Young House, which still stands on Cornell Road and is being renovated to become a Tualatin Hills Park & Recreation District facility.

Geography

According to the United States Census Bureau, the CDP has a total area of 3.7 square miles (9.6 km^2), all of it land.[3]

Demographics

As of the census[1] of 2000, there were 12,597 people, 4,723 households, and 3,428 families residing in the CDP. The population density was 3,388.3 people per square mile (1,307.5/km²). There were 4,951 housing units at an average density of 1,331.7 per square mile (513.9/km²). The racial makeup of the CDP was 85.42% White, 0.86% African American, 0.41% Native American, 7.47% Asian, 0.21% Pacific Islander, 2.70% from other races, and 2.94% from two or more races. Hispanic or Latino of any race were 6.09% of the population.

There were 4,723 households out of which 39.4% had children under the age of 18 living with them, 60.8% were married couples living together, 8.3% had a female householder with no husband present, and 27.4% were non-families. 21.0% of all households were made up of individuals and 4.3% had someone living alone who was 65 years of age or older. The average household size was 2.67 and the average family size was 3.12.

In the CDP the population was spread out with 28.4% under the age of 18, 6.9% from 18 to 24, 31.1% from 25 to 44, 25.5% from 45 to 64, and 8.1% who were 65 years of age or older. The median age was 36 years. For every 100 females there were 99.7 males. For every 100 females age 18 and over, there were 97.0 males.

The median income for a household in the CDP was $65,730, and the median income for a family was $79,529. Males had a median income of $62,901 versus $36,369 for females. The per capita income for the CDP was $33,555. About 4.7% of families and 6.7% of the population were below the poverty line, including 7.4% of those under age 18 and 5.0% of those age 65 or over.

Infrastructure

A non-profit organization named the Cedar Mill Community Library Association opened the area's first public library in 1976, located in a shopping center at the intersection of Cornell Road and Saltzman Road. Following expansions in 1978, 1988 and 2001, the Cedar Mill Community Library now occupies 24500 square feet (2280 m^2) of space.[4] As of late 2011, on Cedar Mill's southern border, work is nearing completion on several miles of US 26, increasing from 2 to 3 lanes in each direction.

See also

- Cedar Hills, Oregon – nearby area, with a similar name

References

[1] "American FactFinder" (http://factfinder.census.gov). United States Census Bureau. . Retrieved 2008-01-31.
[2] "US Board on Geographic Names" (http://geonames.usgs.gov). United States Geological Survey. 2007-10-25. . Retrieved 2008-01-31.
[3] "US Gazetteer files: 2010, 2000, and 1990" (http://www.census.gov/geo/www/gazetteer/gazette.html). United States Census Bureau. 2011-02-12. . Retrieved 2011-04-23.
[4] "About Cedar Mill Community Library" (http://www.wccls.org/your_libraries/detail.cfm?lc=CML). Washington County Cooperative Library Services. . Retrieved 2009-05-13.

External links

- Cedar Mill community website (http://cedarmill.org/)
- Cedar Mill Library website (http://library.cedarmill.org)

Washington County, Oregon

Washington County, Oregon

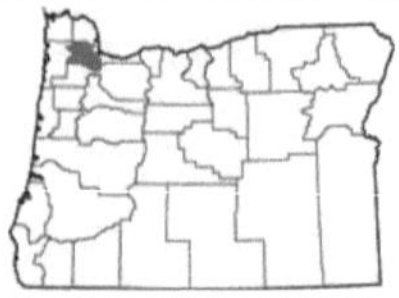

Location in the state of Oregon

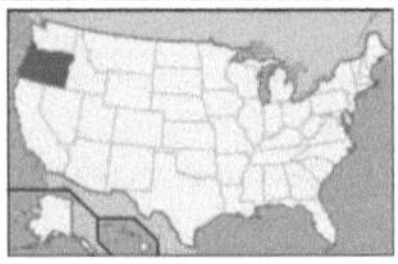

Oregon's location in the U.S.

Founded	September 3, 1849
Seat	Hillsboro
Area **- Total** **- Land** **- Water**	726 sq mi (1880 km²) 724 sq mi (1875 km²) 3 sq mi (8 km²), 0.36%
Population **- (2010)** **- Density**	529710 732/sq mi (282.5/km²)
Website	www.co.washington.or.us [1]

Washington County is one of 36 counties in the U.S. state of Oregon. Originally named **Twality** when created in 1843, the territorial legislature renamed it for the first president of the United States, George Washington, in 1849. The original boundaries included the entire northwest corner of Oregon before sections became new counties. Washington County contains only one river, the Tualatin River, and its drainage basin is almost entirely within the county. The county is nearly coterminous with the Tualatin Valley and is bordered on the west and north by the Northern Oregon Coast Range, on the south by the Chehalem Mountains, and on the north and east by the Tualatin Mountains (or West Hills).

In the 2010 census, its population was 529,710, making it the second most populous county in the state.[2] The seat and largest city is Hillsboro.[2] Other cities include Beaverton, Tigard, and Forest Grove, the oldest city in the county.[3] Major roads in the county include small sections of Interstate 5 and Interstate 205, the Sunset Highway, Oregon Route 217, Oregon Route 47, Oregon Route 10, Oregon Route 6, and Oregon Route 8. Public transportation is primarily operated by TriMet and includes buses, the Westside Express Service commuter rail, and MAX Light Rail. Other transportation includes air travel at the Hillsboro Airport, several private airfields and heliports, and heavy rail cargo on several rail lines.

Government

The county is governed by an elected board of five commissioners. The county is divided into four commissioner districts. One commissioner sits for each district, and the fifth commissioner is "at large" and is the Chair of the board.[4]

Politics

Washington County generally leans Democratic but can be closely divided. In 2008, Barack Obama won with 59.8% of the popular vote (141,544 total votes) to Republican John McCain's with 37.7% (89,185). For the Senatorial race that year, it was much closer with Democrat Jeff Merkley winning 48.8% of the vote (111,367); Republican Gordon H. Smith won 46.5% of the vote (106,114).[5]

Economy

The Washington County Courthouse in Hillsboro

Washington County is centered on a fertile plain that attracted farmers before the first wagon trains. In 1997, orchards covered 8,403 acres (34 km²) of the county's lands and 1,163 acres (4.7 km²) were devoted to vineyards. Agriculture is still a major industry in Washington County, as are lumber, manufacturing, and food processing. (The county is home to Roloff Farms, featured in the TV show *Little People, Big World.*)

The development of a large electronics industry during the 1980s and 1990s is the dominating factor of the county economy. California-based Intel, Oregon's largest private for-profit employer, has its largest concentration of employees in the county, mainly in Hillsboro. Other technology companies include Electro Scientific Industries, Lattice Semiconductor, FEI Company, Merix, TriQuint Semiconductor, Tektronix, SolarWorld, Planar Systems, and EPSON.

Nike, one of two Fortune 500 corporations based in Oregon, has its headquarters in Washington County. Until it was acquired by IBM, Sequent Computer Systems was headquartered right next door to Nike. The facility is now host to a number of software groups for IBM, including one of its Linux Technology Centers. Other companies with headquarters in Washington County include optical instruments manufacturer Leupold & Stevens, Columbia Sportswear, and Reser's Fine Foods.

Geography

Located approximately 20 miles (32 km) to the west of Portland, the county has a total area of 726 square miles (1,881 km²), of which 724 square miles (1,874 km²) is land and 3 square miles (7 km²) (0.36%) is water. The Portland Metro Urban Growth Boundary (UGB) bisects the county. The county's highest point is South Saddle Mountain at 3,464 feet (1,056 m) above sea level in the Northern Oregon Coast Range.[6]

Most of the county is in the Tualatin Valley, which is surrounded by the Tualatin Mountains to the east and north, the Chehalem Mountains to the south, and the Northern Oregon Coast Range to the west and north. In the valley is the only river in the county, the Tualatin River, the Tualatin Plains, and several small hills. The northern and western portions of the county are forested, while the remainder of the county includes urban areas, agricultural lands, and floodplains.

Adjacent counties

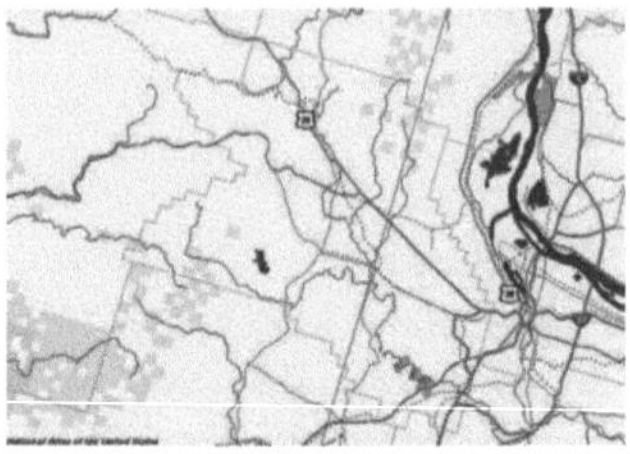
Map of Washington County

Washington County is one of five Oregon counties making up the Portland metropolitan area. The other four are:

- Multnomah County, Oregon - (east)
- Clackamas County, Oregon - (southeast)
- Columbia County, Oregon - (north)
- Yamhill County, Oregon - (south)

Adjacent rural counties:

- Tillamook County, Oregon - (west)
- Clatsop County, Oregon - (northwest)

Major highways

- Interstate 5
- Interstate 205
- U.S. Route 26
- Oregon Route 6
- Oregon Route 8
- Oregon Route 10
- Oregon Route 47
- Oregon Route 99W
- Oregon Route 210
- Oregon Route 217
- Oregon Route 219

National protected area

- Tualatin River National Wildlife Refuge (part)

Demographics

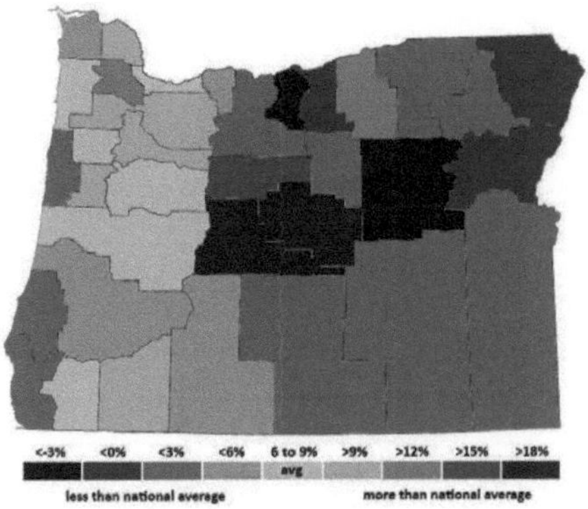

From 2000 to 2007, Washington County's population growth was 14.8%, twice the national average. It was the fastest growing county in the Portland metro area.

Historical populations		
Census	Pop.	%±
1850	2652	—
1860	2801	5.6%
1870	4261	52.1%
1880	7082	66.2%
1890	11972	69.0%
1900	14467	20.8%
1910	21522	48.8%
1920	26376	22.6%
1930	30275	14.8%
1940	39194	29.5%
1950	61269	56.3%
1960	92237	50.5%
1970	157920	71.2%
1980	245808	55.7%
1990	311554	26.7%
2000	445342	42.9%
2010	529710	18.9%
sources:[7] [8] [9]		

As of the census[10] of 2000, there were 445,342 people, 169,162 households, and 114,015 families residing in the county. The population density was 615 people per square mile (238/km²). There were 178,913 housing units at an average density of 247 per square mile (95/km²). The racial makeup of the county was 82.19% White, 1.15% Black or African American, 0.65% Native American, 6.68% Asian, 0.30% Pacific Islander, 5.86% from other races, and 3.17% from two or more races. 11.17% of the population were Hispanic or Latino of any race. 17.2% were of German, 9.9% English, 8.2% Irish and 6.7% American ancestry according to Census 2000. 81.7% spoke English, 9.6% Spanish and 1.2% Vietnamese as their first language.

There were 169,162 households out of which 35.60% had children under the age of 18 living with them, 54.50% were married couples living together, 9.00% had a female householder with no husband present, and 32.60% were non-families. 24.70% of all households were made up of individuals and 6.70% had someone living alone who was 65 years of age or older. The average household size was 2.61 and the average family size was 3.14.

In the county, the population was spread out with 26.90% under the age of 18, 9.30% from 18 to 24, 34.10% from 25 to 44, 20.90% from 45 to 64, and 8.80% who were 65 years of age or older. The median age was 33 years. For every 100 females there were 99.10 males. For every 100 females age 18 and over, there were 97.00 males.

The median income for a household in the county was $52,122, and the median income for a family was $61,499. Males had a median income of $43,304 versus $31,074 for females. The per capita income for the county was $24,969. About 4.90% of families and 7.40% of the population were below the poverty line, including 8.30% of those under age 18 and 5.30% of those age 65 or over.

History

The Provisional Legislature of Oregon created the county as Twality District on July 5, 1843.[11] Twality was one of the original four districts of the Provisional Government of Oregon in Oregon Country along with Clackamas, Champooick (later Marion), and Yamhill counties. Columbia, later known as Hillsboro, was selected as the county seat in 1850. Washington County obtained its present boundaries in 1854 with the creation of Columbia County to the north and Multnomah County to the east.[11]

Current county jail in Hillsboro.

The construction of Canyon Road to Beaverton helped Portland to consolidate its position as the primary port of Oregon, and defeat the rival efforts of settlements as Oregon City and Milwaukie.

In November 2004, the County and the City of Beaverton agreed to a plan where the city would annex both unincorporated residential neighborhoods as well as high-value areas of land. This would result with Cedar Hills, Garden Home, Raleigh Hills, West Slope being incorporated by 2010, and the communities of Aloha, Bethany, and Cedar Mill at some point after that.

Those plans have since been put on hold since after the City of Beaverton attempted to forcibly annex Nike, Inc.'s World Headquarters, which would have increased Nike's taxes substantially. Nike successfully lobbied for a law that would prohibit their annexation for 99 years. Since that decision, annexation plans have been halted and Washington County has started urban planning to provide city-level services to the unincorporated urban areas in the county.[12]

See also

- David Hill
- Tabitha Brown
- Joseph Meek
- Washington County Jail
- Washington County Courthouse

Communities

Incorporated cities

Downtown Beaverton

Downtown Forest Grove in 1920

City	1990 population	2000 population	2010 population[13]	Incorporated	Notes
Banks	563	1,286	1,777	1921	
Beaverton	53,310	76,129	89,803	1893	
Cornelius	6,148	9,652	11,869	1893	
Durham	748	1,382	1,351	1966	
Forest Grove	13,559	17,708	21,083	1872	
Gaston	563	600	637	1914	
Hillsboro	37,520	70,187	91,611	1876	County seat
King City	2,060	1,949	3,111	1966	
Lake Oswego	30,576	35,278	36,619	1910	Small portion, most in Clackamas County[14]
North Plains	972	1,605	1,947	1963	
Portland	437,319	529,121	583,776	1851	Small portion, most in Multnomah County[15]
Rivergrove	294	324	289	1971	Small portion, most in Clackamas County
Sherwood	3,093	11,791	18,194	1893	
Tigard	29,344	41,223	48,035	1961	
Tualatin	15,013	22,791	26,054	1913	Small portion also in Clackamas County
Wilsonville	7,106	13,991	19,509	1969	Small portion, most in Clackamas County[16]

Unincorporated communities and CDPs

- Aloha
- Bacona
- Balm Grove
- Bethany
- Blooming
- Bonita
- Bonny Slope
- Bradley Corner
- Buckheaven
- Bull Mountain
- Buxton
- Carnation
- Cedar Hills
- Cedar Mill
- Chehalem
- Cherry Grove
- Dilley
- Elmonica
- Farmington
- Gales Creek
- Garden Home
- Glencoe
- Glenwood
- Hayward
- Hazeldale
- Helvetia
- Hillside
- Huber
- Kansas City
- Kinton
- Laurel
- Laurelwood
- Manning
- Marlene Village
- Metzger
- Middleton
- Midway
- Mountaindale
- Mulloy
- Norwood
- Oak Hills
- Orenco
- Patton
- Progress
- Raleigh Hills
- Reedville

- Rockcreek
- Roy
- Scholls
- Six Corners
- Somerset West
- Tanasbourne
- Thatcher
- Timber
- Tobias
- Verboort
- Watts
- West Haven
- West Slope
- West Union
- Whitford
- Wilkesboro
- Witch Hazel

See also

- National Register of Historic Places listings in Washington County, Oregon
- Washington County Museum
- L. L. "Stub" Stewart State Park
- Ki-a-Kuts Falls

References

[1] http://www.co.washington.or.us

[2] Proehl, Risa S. (March 2009). "2008 Oregon Population Report" (http://www.pdx.edu/sites/www.pdx.edu.prc/files/media_assets/PopRpt08c2.pdf). *Population Research Center*. Portland State University. pp. 7. . Retrieved 2009-05-05.

[3] "Incorporation dates". *The Hillsboro Argus*: pp. Communities, p. 21. October 19, 1976.

[4] Official County web site (http://www.co.washington.or.us/cgi/home/washco.pl)

[5] http://uselectionatlas.org/RESULTS/

[6] Peakbagger.com: South Saddle Mountain, Oregon (http://www.peakbagger.com/peak.aspx?pid=1076)

[7] census.gov Oregon population by county, 1900-90 (http://www.census.gov/population/cencounts/or190090.txt) - accessed 2009-05-02

[8] quickfacts.census.gov - Washington County (http://quickfacts.census.gov/qfd/states/41/41067.html) - accessed 2010-06-18

[9] http://factfinder2.census.gov

[10] "American FactFinder" (http://factfinder.census.gov). United States Census Bureau. . Retrieved 2008-01-31.

[11] Oregon State Archives: Washington County History (http://arcweb.sos.state.or.us/county/cpwashingtonhome.html)

[12] http://blog.oregonlive.com/oregonianextra/2006/06/appellate_court_rejects_beaver.html

[13] "2010 Census Redistricting Data (Public Law 94-171) Summary File" (http://factfinder2.census.gov/faces/tableservices/jsf/pages/productview.xhtml?pid=DEC_10_PL_GCTPL2.ST13&prodType=table). *American FactFinder*. United States Census Bureau. . Retrieved August 12, 2011.

[14] City of Lake Oswego (http://www.ci.oswego.or.us/engineer/maps/LO_CompPlan.pdf)

[15] PortlandOnline (http://www.portlandonline.com/shared/cfm/image.cfm?id=51672)

[16] City of Wilsonville (http://www.ci.wilsonville.or.us/Index.aspx?page=466)

External links

- Washington County Visitors Association (http://www.wcva.org)

Lake Oswego, Oregon

Lake Oswego, Oregon

— City —

Oswego Lake is in the center of town.

Flag

Seal

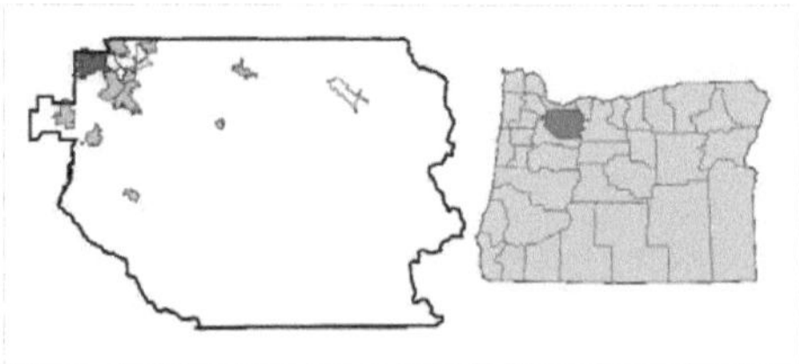

Location in Oregon

Coordinates (City Hall): 45°25′10″N 122°40′03″W

Country	United States
State	Oregon
Counties	Clackamas, Multnomah, Washington
Founded	1847, incorporated 1910
Government	

• Mayor	Jack Hoffman
• City Council	Jeff Gudman Donna Jordan Mike Kehoe Sally Moncrieff Mary Olson Bill Tierney
Area	
• Total	11.3 sq mi (29.3 km^2)
• Land	10 sq mi (25.9 km^2)
• Water	1.3 sq mi (3.4 km^2)
Elevation	sea level to 972 ft (0 to 296 m)
Population (2010)	
• Total	36,619
• Density	3409.7/sq mi (1316.5/km^2)
Time zone	Pacific (UTC-8)
• Summer (DST)	Pacific (UTC-7)
ZIP codes	97034-97035
Area code(s)	503 and 971
FIPS code	41-40550[1]
GNIS feature ID	1166669[2]
Website	www.ci.oswego.or.us [3]

Lake Oswego (/ɒsˈwiːɡoʊ/) is a city located primarily in Clackamas County in the U.S. state of Oregon. Small portions of the city are also located in neighboring Multnomah and Washington counties.[4] Located south of Portland surrounding the 405-acre (1.64 km^2) Oswego Lake, the town was founded in 1847 and incorporated as **Oswego** in 1910. The city was the hub of Oregon's brief iron industry in the late 19th century and is today an affluent suburb of Portland. As of the 2010 census, the city had a total population of 36,619,[5] a 3.8% increase over the 2000 population of 35,278.

History

Early history

The Clackamas Indians once occupied the land that later became Lake Oswego,[6] but diseases transmitted by European explorers and traders killed most of the natives. Before the influx of non-native people via the Oregon Trail, the area between the Willamette River and Tualatin River had a scattering of early pioneer homesteads and farms.

19th century

As settlers arrived, encouraged by the Donation Land Claim Act of 1850 and the subsequent Homestead Act, they found the land under-occupied.

Albert Alonzo Durham founded the town of **Oswego** in 1847, naming it after Oswego, New York.[7] He built a sawmill on Sucker Creek (now Oswego Creek), the town's first industry.[6]

In 1855, the federal government forcibly relocated the remaining Clackamas Indians to the Grand Ronde Indian Reservation in nearby Yamhill County.[6]

During this early period in Oregon history, most trade proceeded from Portland to Oregon City via the Willamette River, and up the Tualatin River Valley through Tualatin, Scholls, and Hillsboro. The thick woods and rain-muddied roads were major obstacles to traveling by land. Along the rivers of this area can still be seen the vestiges of river landings, ferry stops, and covered bridges of this period. A landing in the city's present-day George Rogers Park is thought to have been developed by Durham around 1850 for lumber transport; another landing was near the Tryon Creek outlet into the Willamette.

Restored remains of the 1866 Oregon Iron Company furnace, George Rogers Park.

In 1865, prompted by the earlier discovery of iron ore in the Tualatin Valley, the Oregon Iron Company was incorporated. Within two years the first blast furnace on the west coast was built, patterned after the arched furnaces common in northwestern Connecticut, and the company set out to make Oswego into the *Pittsburgh of the West.*[8] In 1878, the company was sold off to out-of-state owners and renamed the Oswego Iron Company, and in 1882, Portland financiers Simeon Gannett Reed and Henry Villard purchased the business and renamed it the Oregon Iron and Steel Company.[9]

The railroad arrived in Oswego in 1886, in the form of the Portland and Willamette Valley Railway. A seven-mile-long line provided Oswego with a direct link to Portland. Prior to this, access to the town was limited to primitive roads and river boats. The railroad's arrival was a mixed blessing; locally, it promoted residential development along its path, which enabled Oswego to grow beyond its industrial roots. But nationally, the continued expansion of freight railroad system gave easy local access to cheaper and higher quality iron from the Great Lakes region. This ultimately led to the local industry's demise.[6] [9]

By 1890, the industry produced 12,305 tons of pig iron, and at its peak provided employment to around 300 men. The success of this industry greatly stimulated the development of Oswego, which by this time had four general stores, a bank, two barber shops, two hotels, three churches, nine saloons, a drugstore, and even an opera house.[8]

The iron industry was a vital part of a strategy designed by a few Portland financiers who strove to control all related entrepreneurial ventures in the late 19th century. Control of shipping and railroads was held under the Oregon Steam Navigation Company, later to become the Oregon Railway and Navigation Company. This local monopoly responded to the area's increasing demand for iron and steel, and grew to play a key role in economic history throughout the area.

20th century

The Oregon Iron and Steel Company adapted to the new century by undertaking programs in land development, selling large tracts of the 24000 acres (97 km^2) it owned, and power, building a plant on Oswego Creek starting in 1905, and erecting power poles in subsequent years to supply power to Oswego citizens. With the water needs of the smelters tailing off, the recreational potential of the lake and town was freed to develop rapidly.[6]

Lake Oswego railroad bridge across the Willamette River, Lake Oswego, Oregon, April 2008

In 1910, the town of Oswego was incorporated.[6] The Southern Pacific Railroad, which had acquired the P&WVR line at the end of the 19th century, widened it from narrow to standard gauge and in 1914, electrified it, providing rapid, clean, and quiet service between Oswego and Portland. The service was known as the Red Electric.[6]

Passenger traffic hit its peak in 1920 with 64 trains to and from Portland daily. Within nine years of the peak, passenger service ended and the line was used for intermittent freight service to Portland's south waterfront up until its abandonment in 1984. The line was preserved, however and the Willamette Shore Trolley provides tourist rides on the line today.

One of the land developers benefiting from sales by OI&S was Paul Murphy, whose Oswego Lake Country Club helped promote the new city as a place to "live where you play."[6] Murphy was instrumental in developing the first water system to supply the western reaches of the city, and also played a key role in encouraging the design of fine homes in the 1930s and 1940s that ultimately would establish Oswego as an attractive place to live. In the 1940s and 1950s, continued development helped spread Oswego's residential areas.[6]

Mass transit service after the end of electric interurban service was provided by Oregon Motor Stages, but that company suspended all operations following a drivers strike in 1954.[10] In 1955, a newly formed private company, Intercity Buses, Inc., began operating bus service connecting Oswego with downtown Portland and Oregon City.[11] This service was taken over by TriMet in 1970.

In 1960, Oswego was renamed Lake Oswego when it annexed part of neighboring Lake Grove.[6]

Geography

According to the United States Census Bureau, the city has a total area of 10.9 square miles (28.2 km^2), of which 10.4 square miles (26.9 km^2) is land and 0.6 square miles (1.6 km^2) is water. That area does not include the more than 1100 acres (4.5 km^2) of unincorporated land within the urban services boundary as defined by Clackamas County.[5]

Oswego Lake is a private lake (an expansion of an earlier natural lake, named *Waluga* (wild swan) by Clackamas Indians[12]) managed by the Lake Oswego Corporation.[13] Houses with views of Oswego Lake and lake-shore property are very expensive. The lake is navigable, with a dock at the east end where boaters can disembark and walk to the nearby businesses. The main canal from the Tualatin River was dug in the early part of the century.

Every five to ten years, the water level in the lake is lowered several feet by opening the gates on the dam and allowing water to flow into Oswego Creek and on to the Willamette River, enabling lakefront property owners to conduct repairs on docks and boathouses.[14] In 2010, the lake was lowered approximately 24 feet (7.3 m) to allow for construction of a new sewer line, the lowest lake level since 1962 when the original sewer line was installed.[15]

The city extends up Mount Sylvania and through Lake Grove towards Tualatin.

Demographics

Historical populations		
Census	Pop.	%±
1920	1818	—
1930	1285	-29.3%
1940	1726	34.3%
1950	3316	92.1%
1960	8906	168.6%
1970	14573	63.6%
1980	22527	54.6%
1990	30576	35.7%
2000	35278	15.4%
2010	36619	3.8%
U.S. Decennial Census [16]		

Lake Oswego is one of the most affluent suburbs of Portland. In 2000, the city had a median household income of $71,597, up from $57,499 in 1990. Additionally, as in the rest of the Portland metropolitan area, house prices have increased rapidly (as of June 2006); the median value in 2000 was $296,200, over twice what it was in 1990 ($142,600).

2000 Census data

As of the census of 2000, there are 35,278 people in the city, organized into 14,769 households, and 9,665 families.[1] The population density is 3409.7 sq mi (8831 km^2). There are 15,741 housing units at an average density of 1,521.4 per square mile (587.2/km²). The racial makeup of the city is 91.1% White, 4.6% Asian, 0.6% Black or African American, 0.3% Native American, 0.2% Pacific Islander, 0.7% from other races, and 2.5% from two or more races. 2.3% of the population are Hispanic or Latino of any race. There are 14,769 households out of which 32.0% have children under the age of 18 living with them, 56.2% are married couples living together, 6.9% have a female householder with no husband present, and 34.6% are non-families. 27.9% of all households are made up of individuals and 7.9% have someone living alone who is 65 years of age or older. The average household size is 2.38 and the average family size is 2.95.

In the city the population dispersal was 24.8% under the age of 18, 6.1% from 18 to 24, 26.8% from 25 to 44, 31.0% from 45 to 64, and 11.4% who are 65 years of age or older. The median age is 41 years. For every 100 females there are 92.9 males. For every 100 females age 18 and over, there are 88.2 males. The median income for a household in the city is $71,597, and the median income for a family is $94,587 (These figures had risen to $76,883 and $105,763 respectively as of a 2007 estimate.)[17] Males have a median income of $66,380 versus $41,038 for females. The per capita income for the city is $42,166. 3.4% of the population and 2.3% of families are below the poverty line. Out of the total population, 2.0% of those under the age of 18 and 4.0% of those 65 and older are living below the poverty line.

City government

The city has a council-manager form of government, which vests policy-making authority in an elected, volunteer city council. The council consists of a mayor and six councilors, all of whom are elected at-large and serve four-year terms.

Day-to-day operations are handled by an appointed, professional city manager. Almost all of the city's employees, which include part-time staff amounting to approximately 342 full-time equivalents, report to the city manager. This includes the police chief, fire chief, one assistant city manager, and the community development director. The biggest groups are:

- police and fire departments, consisting of about 50 people each,
- the library, parks, and recreation departments, consisting of about 70 people total, and
- about 80 people throughout the engineering, planning, and maintenance departments.

Civic involvement

Neighborhood associations play a formal role for citizen involvement in the city government's land-use planning and other activities. A neighborhood association's role is governed by state and city law. As of February 2005, there are 20 recognized neighborhood associations: Blue Heron ¤, Bryant ¤, Country Club-North Shore ¤, Evergreen ¤, First Addition, Forest Highlands, Glenmorrie, Hallinan, Holly Orchard, Lake Forest, Lake Grove, Lakewood ¤, McVey-South Shore ¤, Oak Creek, Old Town, Palisades ¤, Rosewood, Uplands, Waluga, and Westridge.[18] (Associations including lakefront property are marked with a ¤ symbol.)

The Mountain Park neighborhood, located between Oak Creek and Holly Orchard on its west side and Forest Highlands and Uplands on its east, has a homeowners' association serving the role of neighborhood association for its residents.

The following statistics summarize other aspects of civic involvement:

- Registered voters: 23,061
- Citizen advisory boards: 10
- Community volunteers: 500 plus
- Citizens trained for community emergency response teams: 487

Public schools

Lake Oswego High School

The Lake Oswego School District is within the city boundaries and serves the city's approximately 7,000 students, with a ratio of 23 students per instructor. The two high schools in the district are Lake Oswego High School and Lakeridge High School. There are nine elementary schools and two junior high schools for students in grade 7 and 8.[19]

Lakeridge High School.

Cultural and recreational facilities

The city maintains 573 acres (2.32 km^2) of parks and open spaces. This includes 24 developed parks, one amphitheater, one swim park, one water sports center on the Willamette River, an adult community (senior) center, a public golf course (self-financed), an indoor tennis center, seven outdoor tennis courts, and five picnic shelters. An additional swim park is open to those who live in the boundaries of the old Lake Grove School District. The private Oswego Lake Country Club and neighboring equestrian riding club add to the recreational amenities of the city.

Lake Oswego has one public library, part of the Library Information Network of Clackamas County. For the past six years, the library has been rated among the top 10 libraries serving similar population sizes in the United States.[20]

Notable people

This is a partial list of notable residents, past and present, of Lake Oswego:

- LaMarcus Aldridge (1985–), NBA player for the Portland Trail Blazers[21]
- Allen Alley (1954–), Founder of Pixelworks, Republican nominee for Oregon State Treasurer in 2008, Republican candidate for Oregon Governor in 2010
- Jon Arnett, NFL player and member of the College Football Hall of Fame[22]
- Daniel Baldwin (1960–), film actor, producer, and director[23]
- J. J. Birden (1965–), NFL wide receiver[24]
- Walter F. Brown (1926–), Navy Commander JAGC, Judge, State Senator, 2004 Presidential candidate for the Socialist Party USA[25]
- Terry Dischinger (1940–), basketball gold medalist in the 1960 Olympics and NBA player from 1962–1973 and orthodontist[26]
- Chris Dudley (1965–), NBA player from 1987–2003 and 2010 Republican nominee for Governor of Oregon[27]
- Mike Dunleavy, Jr. (1980–), NBA player for the Indiana Pacers[28]
- Mike Erickson (1963–), businessman and candidate for U.S. Congress in 2006 and 2008[29]
- Rudy Fernández (1985–), NBA player for the Portland Trail Blazers (2008–2011)[30]
- Stu Inman (1926–2007), co-founder of the Portland Trail Blazers[31]
- Neil Lomax (1959–), NFL quarterback 1981–1988[32]
- Kevin Love (1988–), NBA player for Minnesota Timberwolves and former UCLA Bruin[33]
- Stan Love (1949–), NBA player 1971–1975 and father of Kevin Love[33]
- Merrill A. McPeak (1936–), former U.S.A.F. chief of staff[34]
- Bart Miadich (1976-), Major League Baseball pitcher 2001–2006[35]
- Patrick Mills 1988–), NBA player for the Portland Trail Blazers (2009–)[36]
- Linus Pauling (1901–1994), Winner of two Nobel prizes, in peace and chemistry. Author, educator.[37]
- Julianne Phillips (1960–), model, actress, co-star of 1990s TV series *Sisters*[38]
- Richard Read (1957-), Two-time Pulitzer prize winner, reporter for The Oregonian[39] [40]
- Mike Richardson, co-founder/publisher of Dark Horse Comics[41]
- Don Schollander (1946–) Olympic swimming gold medalist in 1964 and 1968[42]
- Henry Selick (1952) American director and animator: The Nightmare Before Christmas, Coraline[43]
- William Stafford (1914–1993), American poet and U.S. poet laureate 1970–1971[44]
- Drew Stanton (1984–), NFL quarterback for the Detroit Lions[45]

- Michael Stutes (1984–), MLB relief pitcher for the Philadelphia Phillies[46]
- Nathan Farragut Twining (1897–1982), Chairman of the United States Joint Chiefs of Staff, 1957–1960[47]

Sister cities

Lake Oswego has one sister city:

- Yoshikawa, Japan[48]

References

[1] "American FactFinder" (http://factfinder.census.gov). United States Census Bureau. . Retrieved 2008-01-31.
[2] "US Board on Geographic Names" (http://geonames.usgs.gov). United States Geological Survey. 2007-10-25. . Retrieved 2008-01-31.
[3] http://www.ci.oswego.or.us/
[4] http://www.ci.oswego.or.us/engineer/maps/LO_CompPlan.pdf
[5] "City population is up – what will that mean?" (http://www.lakeoswegoreview.com/news/story.php?story_id=130092552348458300). Lake Oswego Review. Mar 24, 2011. . Retrieved 2011-03-27.
[6] "A Brief History" (http://www.ci.oswego.or.us/about-lo/history.htm). City of Lake Oswego website. . Retrieved 2006-12-11.
[7] "Eight Myths Concerning Lake Oswego" (http://www.oswegoheritage.org/history/mythsmuddles.html). Oswego Heritage Council. . Retrieved 2010-01-15.
[8] Goodall, Mary (1958). *Oregon's Iron Dream*. Portland, Oregon: Binsford & Mort. p. 43.
[9] Kuo, Susanna Campbell. "A Brief History of the Oregon Iron Industry" (http://www.oswegoheritage.org/history/pdf/history.pdf). Oswego Heritage Council. . Retrieved 2009-07-23.
[10] "Petition of Intercity Buses, Inc., Wins Support of Oswego as PUC Hearing Ends". (December 22, 1954). *The Oregonian*, p. 8.
[11] "Oswego Fete Due Bus Line: Regular Service Set Next Monday". (February 3, 1955). *The Oregonian*, p. 8.
[12] http://www.ci.oswego.or.us/Plan/Comp%20Plan/Goal%205_7.pdf
[13] "Lake Oswego Corporation official website" (http://www.lakecorp.com/). . Retrieved 2011-03-27.
[14] Tims, Dana (October 31, 2006). "Drawdown under way to lower Oswego Lake". *The Oregonian*.
[15] Newell, Cliff (July 29, 2010). "Sewer project will make Oswego Lake disappear – briefly" (http://www.portlandtribune.com/news/story.php?story_id=128035388723593500). *The Oregonian*. . Retrieved October 4, 2010.
[16] http://www.census.gov/prod/www/abs/decennial/
[17] "Lake Oswego city, Oregon" (http://factfinder.census.gov/servlet/ACSSAFFFacts?_event=Search&geo_id=16000US5551150&_geoContext=01000US|04000US55|16000US5551150&_street=&_county=lake+oswego&_cityTown=lake+oswego&_state=04000US41&_zip=&_lang=en&_sse=on&ActiveGeoDiv=geoSelect&_useEV=&pctxt=fph&pgsl=160&_submenuId=factsheet_1&ds_name=ACS_2007_3YR_SAFF&_ci_nbr=null&qr_name=null&reg=null:null&_keyword=&_industry=). U.S. Census Bureau. . Retrieved June 22, 2009.
[18] "Planning: Neighborhood Associations" (http://www.ci.oswego.or.us/plan/Neighborhoods/default.htm). City of Lake Oswego website. . Retrieved 2008-04-26.
[19] "About the District" (http://loswego.k12.or.us/general_info/visitor_info/district_about.htm). Lake Oswego School District. . Retrieved 2009-07-23.
[20] "Hennen's American Public Library Ratings" (http://www.haplr-index.com/ratings.html). . Retrieved 2009-08-12.
[21] Quick, Jason (2009-07-01). "'Faith' keeps LaMarcus Aldridge confident he'll stay with Blazers" (http://blog.oregonlive.com/behindblazersbeat/2009/06/rip_city_revival_faith_keeps_l.html). *The Oregonian*. . Retrieved 2009-07-01.
[22] Newell, Cliff. " Jaguar Jon (http://www.lakeoswegoreview.com/sports/story.php?story_id=120172955066447200)." *The Lake Oswego Review*. January 31, 2008. Retrieved on January 28, 2009.
[23] Bella, Rick (July 14, 2011). "Daniel Baldwin says volatile relationship grew worse after move to Lake Oswego" (http://www.oregonlive.com/lake-oswego/index.ssf/2011/07/daniel_baldwin_gives_television_interview_expressing_fear_for_his_childrens_safety_around_jailed_wif.html). *The Oregonian*. . Retrieved July 15, 2011.
[24] Newell, Cliff. " Birden sees plenty of success after pro football career (http://www.lakeoswegoreview.com/sports/story.php?story_id=121502256482338500)." *The Lake Oswego Review*. July 3, 2008. Retrieved on January 28, 2009.
[25] Mapes, Jeff (June 7, 2010). "Greens pick former state senator for treasurer, but no one yet for Oregon governor" (http://blog.oregonlive.com/mapesonpolitics/2010/06/greens_nominate_former_state_s.html). *The Oregonian*. . Retrieved June 7, 2010.
[26] Newell, Cliff. " Upward Basketball teaches b-ball to youngsters (http://www.lakeoswegoreview.com/sports/story.php?story_id=120596583109906300)." *The Lake Oswego Review*. March 20, 2008. Retrieved on January 28, 2009.
[27] "Dudley jumps into Oregon governor's race" (http://www.lakeoswegoreview.com/news/story.php?story_id=126099263754883000). *Lake Oswego Review*. December 16, 2009. . Retrieved December 16, 2009.
[28] " Dunleavy joins Duke exodus (http://www.usatoday.com/sports/college/basketball/men/stories/2002-05-10-dunleavy-nba.htm)." *USA Today*. May 11, 2002. Retrieved on February 6, 2009.

[29] Law, Steve. " Erickson takes GOP primary in 5th District (http://www.lakeoswegoreview.com/news/story.php?story_id=121132760310304700)." *Lake Oswego Review*. May 20, 2008. Retrieved on February 6, 2009.
[30] Eggers, Kerry. " Rudy excites fans and teammates (http://www.portlandtribune.com/sports/story.php?story_id=123007451572027500)." *Portland Tribune*. January 4, 2009. Retrieved on February 6, 2009.
[31] " OBITUARIES; Stu Inman, 80; helped assemble Portland's NBA champion team (http://pqasb.pqarchiver.com/latimes/access/1207632981.html?dids=1207632981:1207632981&FMT=ABS&FMTS=ABS:FT&date=Feb+01,+2007&author=&pub=Los+Angeles+Times&desc=OBITUARIES;+Stu+Inman,+80;+helped+assemble+Portland's+NBA+champion+team&pqatl=google)." *Los Angeles Times*. February 1, 2007. B9. Retrieved on February 13, 2009.
[32] Moore, Kenny. "This Phone Will Ring On Apr. 28." *Sports Illustrated*. April 27, 1981. 2 (http://vault.sportsillustrated.cnn.com/vault/article/magazine/MAG1124421/2/index.htm). Retrieved on February 13, 2009.
[33] " Stan Love (http://wweek.com/story.php?story=5285)." *Willamette Week*. July 7, 2004. Retrieved on February 13, 2009.
[34] " Lake Oswego general could make Kerry cabinet (http://www.kgw.com/news-local/stories/kgw_091704_news_kerry_cabinet.12920c583.html)." *Associated Press*. Friday September 17, 2004. Retrieved on February 13, 2009.
[35] " Stingers Boost Lead (http://nl.newsbank.com/nl-search/we/Archives?p_product=SLTB&p_theme=sltb&p_action=search&p_maxdocs=200&p_topdoc=1&p_text_direct-0=100E8EB5E5206564&p_field_direct-0=document_id&p_perpage=10&p_sort=YMD_date:D&s_trackval=GooglePM)." *The Salt Lake Tribune*. August 17, 2001. C4. Retrieved on February 13, 2009.
[36] "General Mills" (http://www.portlandtribune.com/sports/story.php?story_id=129970868413944900). Portland Tribune. March 10, 2011. . Retrieved April 20, 2011.
[37] "Eight Myths Concerning Lake Oswego" (http://www.oswegoheritage.org/history/mythsmuddles.html). Oswego Heritage Council. . Retrieved 2011-03-27.
[38] " Personalities (http://pqasb.pqarchiver.com/washingtonpost/access/73621924.html?dids=73621924:73621924&FMT=ABS&FMTS=ABS:FT&type=current&date=Aug+06,+1988&author=&pub=The+Washington+Post+(pre-1997+Fulltext)&desc=PERSONALITIES&pqatl=googlePers)." *The Washington Post*. August 6, 1988. C03. Retrieved on February 13, 2009.
[39] "The Pulitzer Prizes | Citation" (http://www.pulitzer.org/citation/1999-Explanatory-Reporting). Pulitzer.org. . Retrieved 2011-03-27.
[40] "The Pulitzer Prizes | Citation" (http://www.pulitzer.org/citation/2001-Public-Service). Pulitzer.org. . Retrieved 2011-03-27.
[41] Korn, Peter. " It's a fantasy world, probably rooted in reality (http://www.lakeoswegoreview.com/features/story.php?story_id=122048122594453300)." *Lake Oswego Review*. September 4, 2008. Retrieved on February 13, 2009.
[42] " SWIMMERS PACED BY SCHOLLANDER; Oregonian Anchors Relay Team to Gold Medal (http://select.nytimes.com/gst/abstract.html?res=F40F17FD355B1B728DDDAC0994D8415B848AF1D3)." *The New York Times*. Thursday October 15, 1964. Retrieved on February 13, 2009.
[43] Hundhammer, Linda (2009-04-02). "Making Movie Magic" (http://www.westlinntidings.com/features/story.php?story_id=123852178999488300). *WestLinn Tidings* (Pamplin Media Group). . Retrieved 10 March 2010.
[44] " Security, BY WILLIAM STAFFORD (http://pqasb.pqarchiver.com/latimes/access/61555185.html?dids=61555185:61555185&FMT=ABS&FMTS=ABS:FT&type=current&date=Nov+24,+1991&author=&pub=Los+Angeles+Times+(pre-1997+Fulltext)&desc=Security,+BY+WILLIAM+STAFFORD&pqatl=google)." *Los Angeles Times*. November 24, 1991. Book Review. Start Page: 6. Retrieved on February 13, 2009.
[45] Murphy, Austin (October 3, 2005). "A Spartan's Revenge" (http://sportsillustrated.cnn.com/vault/article/magazine/MAG1105340/index.htm). *Sports Illustrated*. . Retrieved December 27, 2009.
[46] http://www.baseball-reference.com/minors/player.cgi?id=stutes001mic
[47] "Cultural Resources Inventory: C.W. Twining House" (http://www.ci.oswego.or.us/plan/Historic_Resources_Advisory_Board/Landmarks/TwiningsHouse/Twinnings.pdf). City of Lake Oswego. . Retrieved 2009-05-21.
[48] "Community: History and Culture" (http://www.ci.oswego.or.us/home/community/hist.htm). City of Lake Oswego website. . Retrieved 2006-12-11.

External links

- Historic photos of Lake Oswego (http://www.ci.oswego.or.us/photos/) from the City of Lake Oswego
- Lake Oswego (http://bluebook.state.or.us/local/cities/lr/lakeoswego.htm) from the Oregon Blue Book
- Lake Oswego (http://www.oregonencyclopedia.org/entry/view/lake_oswego/) in the *Oregon Encyclopedia*

Article Sources and Contributors

Nau (clothing retailer) *Source*: http://en.wikipedia.org/w/index.php?title=Nau_%28clothing_retailer%29 *Contributors*: Aboutmovies, Alvin Lee, Baker.lyon, Coreydu, FSII, Faster4tec, Jfire, Nono64, Slon02, Travis.Thurston, Valfontis, Woohookitty, Xnatedawgx, 5 anonymous edits

Portland, Oregon *Source*: http://en.wikipedia.org/w/index.php?title=Portland%2C_Oregon *Contributors*: 08OceanBeach SD, 0x6D667061, 10Barco, 123username, 187onyouboy, 2004-12-29T22:45Z, 5 albert square, 7265, 7george7, A1%, AWeenieMan, Aaronmz, Abendigoreebs, Aboutmovies, Acalamari, Accurizer, Acmilan15, Acntx, AdamjVogt, Adamschneider, Addshore, Adirnasexual, Admrboltz, Aeranis, Agassiz830, Ahoerstemeier, Ajbenj, Akhilleus, Alantex, Alburns, Ale jrb, Alexclemens, AlexiusHoratius, Allen4names, Alphalife, Amadeust, Amateria1121, Amatulic, Amazonien, Amberroom, Andattaca2010, Andre Engels, Andrearennie, Andrewlp1991, Andrewpmk, Andy Marchbanks, Andycjp, Animum, Ann Stouter, Another Believer, Antandrus, Arbeo, Arch dude, Archer3, Ardizzone, AreJay, Artemisboy, Arturoramos, Arvindn, Arxiloxos, Aspects, AtlasPDX, Avnjay, Axcordion, AxelBoldt, Ayip520, BUF4Life, BalooUrsidae, Banana7070, Bancroft595, Barek, Barte, Bartleby, Baseball Bugs, Bazza 7, Bdk, Bears16, Beaverhead420, Beetstra, Beiber belieber, Belongsinaromancenovel, Bender235, Benjamin93, Benzamin, Bfigura's puppy, Bigcheesebebbs, BillGarrett, Billder, Billmason, Bingo fetherstone, Biruitorul, Bizxman, Blackberrylaw, Blackcats, Blackij, Blaze33541, Bletch, Bloodshedder, Blueboy96, Bluenote117, Bobblehead, Bobblewik, BokicaK, Bolivian Unicyclist, Bonneria, Born2cycle, Boy asunder, Braaropolis, BrendtHess, Brian Pearson, Brianga, Brideshead, Brighterorange, Brookie, Bruinfan12, Bryan the Magnificent, Bsskat5, Buchanan-Hermit, Buddha24, Buisychild, Bumm13, Bunchofgrapes, BurningCandy, Butwhatdoiknow, Bvandervoo, CALR, CCookkiieeSS, CGorman, CJ, CJLippert, Cacophony, Cactus.man, Calisafe, CalmCalamity, Caltas, Camembert, Campbead, Can't sleep, clown will eat me, CanadianLinuxUser, Candle421, Cargoking, Carnac32, Carrite, Cartermlb, Catinator, Cburnett, Ccj, Cdc, CecilyBug, Ceyockey, Cfallin, Chadnewt, Charlesemorganiv, CharlotteWebb, Charmpop, Chawman69, Chick Bowen, Chickaoww1234, Chira, Chowbok, Chr.K., Chris the speller, Christopher Mann McKay, Christy747, Chriswaterguy, Chromakode, Chuckgeorge60, Ckatz, Clconway, Clintheacock66, Clipdude, ClydeGlide, Cmputer, Cnnw@cnnw.com, Codysamglaze, Coldmachine, Coldpaws, Colonies Chris, CommonsDelinker, Compassion Info, Conversion script, Cool Stuff Is Cool, Cordell, Coulee, Countrokula, Coyote42, Creidieki, Criticalthinker, Crojik, Cromwellt, Crosstrips, Cst17, Cultureshockpdx, Cultureshok, Curps, Cutlass Supreme, Cww, Cyrusc, DBQer, DEshbaugh, DMWishnow, DVD R W, Da illest, Dale Arnett, Dalit Llama, Daltnpapi4u, Dalydaly, Dan100, Daniel Olsen, Danielsarasota, Danreitz, Darkpool, Darlene4, Darloftus, Darwinek, Davehi1, David Gale, David Levy, Davidphogan74, Davidzundel, Davydog, Dawnseeker2000, Dbc12, Ddon, Decumanus, Delirium, DeltaQuad.alt, Demographia, DemonJuice, DerBorg, DerRichter, Derek.cashman, Devolution, DiJay, Diderot's dreams, Dina, Dino, Discospinster, Dismas, Disneywizard, Divinus, Dmendell, Dmsar, Docu, Donzilla24, Doopokko, Doyoao, Dr. Blofeld, Dr. Zaret, Dralwik, Drmies, Droll, Dryazan, DubbaG, Duck25, Duecejuices, Dyedblonde, Dynaflow, Eaefremov, Ealasaid, Ebeaverton, Eco84, Edward, Edwardsully, Ehurtley, Ekips, Ekotkie, Elena, Eliah, EncMstr, Endomion, Energy Dome, EngineerScotty, EnricoV, Epbr123, Ericl234, Erik9, Esprqii, Ev, Everyking, Excirial, Extraeme, Eyeballjim, Eyrian, Factsbefactsyo, Falldownthestairs, Faustus37, Fcb981, Feedle, FeiTeng1000, Fezzak, Ffaillace, Fg2, Finetooth, Firsfron, Flatface, Flightx52, Foogus, Frania Wisniewska, Frannie.dd, Freddy8070, FreedomCultist, FreplySpang, Froggy52, Fukewe, Fullerton, Fumb1e, Funnyhat, Fusillijerry86, Fuzzy510, Fxer, Félix, GTBacchus, Gadfium, Gaius Cornelius, Gary King, Gcavener, Gdo01, Gezzas Man, Ghoulinator, Gilgamesh, Gkleinman, Gobonobo, Godfrey Daniel, Gogo Dodo, Goldfishbutt, Goldminecouture, Googuse, Graham Jones, Greenshed, GregorB, Greysharbor, Ground Zero, Grumpyyoungman01, Gtrmp, Hadal, Haham hanuka, HangingCurve, Happywaffle, Harkathmaker, Haunted Happenings, Hawk00eyed, Hayaji, Hazel77, Hdt83, Hehehe23456, Helvetica, HenryLi, Henryodell, HexaChord, Hiro Antagonist, HockeyLovingHoser, Houlding, Howcheng, Hubertfarnsworth, Huhwaitwhoareyou, Husond, Hux, I am Girl, Iamhove, IanMac, IanOsgood, Ianrosenberg1, Icantwait, Icelight, Ichabod, Ihateitwhenpeopletakemyname, Ikluft, InaMaka, Infoeco, Infrogmation, Iridescent, Isaacada1, IsambardBrunel, Ivarercool, Ixfd64, Izno, J-beda, J.delanoy, JForget, JNW, JNards007, Jae2006, JakeVortex, JamesAM, Jamesmusik, JamesofMaine, Jamidwyer, Jason Schlumbohm, Jauerback, Jcook980, Jdblack, JeLuF, Jechtinaw, Jeffq, JeffreyAllen1975, Jeffreykopp, Jennifer elder, Jes5199, Jespdx, JesterCountess, Jettblack777, Jgilhousen, Jguk, Jgwpdx, Jhans6135, Jhendin, JimWae, Jimipants, Jjjjud, Jkwiki83, Jlhw, Jllm06, Jmchuff, Jmtn20, Jnstewart, Joan.ignasi.91, JocobFTB2, Joe.kelly503, Joelotz, Joemeyer, Jogers, John1937, Jolomo, Jonnabuz, Jonxwood, JordanRL, Jordanlund, Josh Parris, Jpaul025, Jsayre64, Judge Trudy, Juliancolton, Jusdafax, K50 Dude, KF, Ka Faraq Gatri, Karlkatzke, Karol Langner, Katr67, Katt, KayteeJC, Kbdank71, Keene1997, KeithH, Kelovy, Khatru2, Khazar, Kherpuppet, KimKnees, Kinema, Kintetsubuffalo, Kitch, Kjk, Kkmd, Kman543210, KnowledgeOfSelf, Koppenlady, Kotra, Krushdiva, Ktr101, Kudzu1, Kukini, L Kensington, LAX, Larry_Sanger, Larrybob, Last1gone, Latics, Laurenrose3091, Lcarsdata, LeaveSleaves, LegalTech, Leonard23, Lestatdelc, Letrappe, LeviathanMist, Levineps, Licor, Liebeaffe, Liface, Lightmouse, Lights, LilHelpa, LinguistAtLarge, Little Otik, Livemixlove, LizClayton, Lizziebuck, Llywrch, Local hero, Lokiloki, Loodog, Lord Pistachio, Lovelymischief, Lucasgreen42, LuckyWizard, Lukobe, Lummysdad, Luna Santin, MECU, MJ94, MJCdetroit, MK8, MPerel, MSGJ, Macnannan, Madison20691, Magog the Ogre, Mainliner.espresso, Makaristos, Mammalandlamb, Maqsarian, Marek69, Mark Arsten, Markoff Chaney, Marshallteller, Marshman, Matatigre36, Mathpianist93, Matt Yeager, MatthewBChambers, Max Schwarz, Maxis ftw, MayberryonMushrooms, Mcmeatm, Meelar, Melchoir, Mentifisto, MescalineBanana, Metal.lunchbox, Metalheadpete, Mhwang86, Micahbrwn, Michellecrisp, Mickproper, Mid-county Memo, Middle angel21, Mifter, Mike Dillon, Mikewiley126, Mindbunny, Miracleworker5263, MisfitToys, Mitico, Mkmcconn, Mlaffs, Mmccalpin, MojaveNC, Moncrief, Moncubus, Monegasque, Monterey Bay, Monty845, Moogletw, Morachnus, More correctly, Morven, Moxfyre, Mr Accountable, MrBoo, MrBread, Mrubio18, Mrwatkins83, Msalt, Multi brad5, Mumbo jiggy, Mumonkan, MuseumTour, Mxn, Myasuda, Mysekurity, Mysid, NE2, NSR, Nankai, Napapjd, Nasnema, Nate Silva, NateDres23, Nathan Johnson, Nathan Ward, Neanderthalprimadonna, Ned Scott, Neko-chan, Neohawk13, Neurophyre, Neutrality, Nev1, Newager, Nichalp, Night Gyr, Nino001, Niteowlneils, NittyG, Nixeagle, Nleamy, Noah Salzman, Noisy, Nolth1st, Nomadictroubadour, Northamerica1000, Northwesterner1, Notapipe, Nsaa, Nwfilm, Nyttend, Obtarska, Odie5533, Off2riorob, Offkilter, Ohconfucius, Ohnoitsjamie, Okiefromokla (old), Omarmobarek07, OnePt618, Ooglehunter, Or3g0n, Oregonianexpert, Oroso, Orphan Wiki, Orthographer, OttoTheFish, OverlappingElvis, Owen, Oxymoron83, P. S. Burton, PDXblazers, Parallax, Parcelbrat, Paris1127, Past-man, Patrick, Paul August, Paul Drye, Paul Erik, Paulr, Penguino13, Penser, Pentawing, Pepso2, Peregrine Fisher, PeteLee, Peteforsyth, Pgk, Phil Bastian, Phil Boswell, PhilKnight, Philip Trueman, Philopp, Philvarner, Piano non troppo, Picklebird, Pingveno, Pinotgris, Pjrich, Pleather, PoccilScript, Pooponastick2.0, Portland events, PortlandGayJew, Portlandium, Postcard Cathy, Postdlf, Prevod&priredba, Pwrs tny, Quietus2Silence, Quuxplusone, Qxz, R'n'B, R4gn4r, RFBailey, RJN, RTucker, Ram-Man, Randy2063, Raven1977, Raven4x4x, Rawstock, Rbedmiston, Rclocher3, Rdsmith4, Realm of Shadows, Redcherrycola, Redvers, Reedie2be, Reedy, Reinthal, Res2216firestar, Resplendent, Reticulum, RevTenderBranson, Rfernand, Rhys, Rich Farmbrough, Richardaedwards, RickK, Rickington, Ricksan37, Rinkidinki, River Jiang, Rjensen, Rjwilmsi, Rkhwaja, Rocket000, Rockingharder, Rocko Billy, RomDolce, Romanrichardinskeep, Romneybeard the Pirate, Rootbeer, Rossdegenstein, Roux, Russer66, Rw rynerson, SDC, SJ Morg, SPUI, SRTrainer, SU Linguist, Sailingfanblues, Saintlovely, Sam8, Samba6566, Sammy1007, Sanfranman59, Sarahandrea, Saravask, SatanicClaw, Scarlet257, Sceptre, Schmerika, Scientizzle, ScottMainwaring, ScottRainey, Scottdoesntknow, Scottrainey, Scottywong, Sean Kelly, Senor Kruzkin, Seth Ilys, Sfdan, Sfmontyo, Sgt Pinback, Shalom Yechiel, Shanezey, ShelfSkewed, Shoeofdeath, Shuish, Shyguy1991, SilkTork, Simulcra, Siruguri, Skibato, Skk646, Sky Attacker, Skybum, Slateface, Slyvan.rls09, Smashville, Smith03, Smith3689, Smokepot, Snuib, SoCal L.A., Someguy1200, Something in the Water, SoundGod3, South Bay, Sowelilitokiemu, SparsityProblem, SpellChecker, Spingirl2009, SpokaneWilly, SportingFlyer, Squidley, Sstigler, Steinsky, Steven Walling, Steviesteve69, Sthomas22, Stick Fig, StradivariusTV, Strickjh2005, Strongbad1982, StuffOfInterest, Stuharu, Sugarcubez, Suntzu3500, Susvolans, Svyatoslav, Sxm475, Synchronism, TEG24601, TJ Spyke, Tabletop, Tapir Terrific, Tassedethe, Tbhotch, Tboy206, Tdferro, Tdogg241, Tea Tzu, Tedder, Teemu Ruskeepää, Template namespace initialisation script, Tepasi, Terryn3, Tesscass, Thaddeus Slamp, The Emirr, The Little Blue Frog, The Thing That Should Not Be, The Universe Is Cool, TheDoctor10, TheOtherJesse, Thefiestysoldier, Thomas Paine1776, Thomasngo, Tibetan Prayer, Tide rolls, TimeClock871, Tiptoety, Titoxd, Tnxman307, Todoeralindo, Tom X. Tobin, TomYears, Tomwsulcer, Tony Sidaway, Tonym88, Tpbradbury, Tradnor, Trashbag, Travis.Thurston, Triona, Triwbe, Trojan51, Trudy Marchany, Truflip99, Trusilver, Trödel, TurabianNights, Tv145033, Twinsday, Twp, Twwalter, Tyrell1234, Tyw7, Uelenator, Ugur Basak, Ukexpat, Ulmanor, Ulric1313, Un sogno modesto, UnqstnableTruth, Unschool, Urbanxplorx, User2004, VQuakr, Valfontis, VanHelsing.16, Vandalism255, VanillaBear23, Vanka5, Vclaw, Vclipper, VegaDark, Vegaswikian, Ventusa, VenusianCat, Vertigo Acid, VeryVerily, ViceEmperor816, Vicki Rosenzweig, Virek, Vishnava, Vroo, Wabernat, WadeSimMiser, Wavelength, Wayward, Weezering, Western Pines, Westernyankee, Westonmr, WhatamIdoing, WhisperToMe, Whisperingbobblehat, White Shadows, Wik, Will Beback, William Gillis, Willking1979, Wjemather, Wknight94, Wmleler, Wokwiki424, Woogee, Woohookitty, WormRunner, WriterHound, Wwoods, Xdenizen, Xiaoyu of Yuxi, Xmike991, Xomgreg, YALEYALE, YBG, Yachtsman1, Yappa, Yk Yk Yk, Yoninah, ZacharyS, Zeashake, Zereshk, Zigger, Zoe, Zonath, Zundark, Zytron, Zzyzx11, 2278 anonymous edits

Mercy Corps *Source*: http://en.wikipedia.org/w/index.php?title=Mercy_Corps *Contributors*: 2toise, AbHa, Aboutmovies, Ahoerstemeier, Ajbenj, Arugam, Bjfg897, Bob98133, Busy Stubber, Busymale, Ccfromcc, DBlomgren, DGG, DavidLevinson, Dawnseeker2000, Debresser, Dzachs, Erika1212, FreplySpang, Froggo Zijgeb, Frymaster, Genuiben, Geohumphrey, Gobonobo, Hu12, Ianblair23, JLaTondre, JTM, Kbdank71, Kgrad, Kingpin13, Kingturtle, Kletta, Kris1490, La goutte de pluie, Larree0, Leah Nevada Page, Lquilter, MDC1975, Macrakis, Mark Richards, Mason Clegg, Mrwojo, Mysid, Nnemo, Ombudsman, Orangemike, Owen, PaulHanson, Pcpcpc, PeachWriter66, Peteforsyth, Petiatil, Pgan002, R'n'B, RedWolf, Richhoncho, RogerBurks, SamJohnston, Scs07, SilyOlPooh, Tassedethe, Two Bananas, Uliv, Wadkdc, Woohookitty, Δ, 70 anonymous edits

Pacific Northwest *Source*: http://en.wikipedia.org/w/index.php?title=Pacific_Northwest *Contributors*: (jarbarf), 08OceanBeach SD, 5 albert square, Aboutmovies, Acdwiki, Ace Fishbone, Adam850, Adamschneider, Ajraddatz, Alansohn, Ale jrb, Allstarecho, Amog, Anaxial, Andy Marchbanks, Angbrown, Angrytoast, Antandrus, Antwonbonson, Arakunem, Arensaini99, AriaShakespeare, Artichoker, AshleyMorton, Avaarga, Avenue, Ayrton Prost, Azdarin, BD2412, BGManofID, Barek, Bazonka, Bball freak 22, Beenturns22, Beepbeepribbyribby, Blackcats, Blehfu, Bobblehead, Bobo192, BrainyBabe, BrianWilloughby, Buchanan-Hermit, Bugo30, CIreland, Cacophony, CalmCalamity, Captain Yankee, Catapult, Causa sui, Cdellwo, Chadlupkes, Ckatz, Clicketyclack, Cmdrjameson, Colonies Chris, Cortesevf, Corvus cornix, Coyote42, CrazyC83, Cuchullain, Dale Arnett, Daniel C. Boyer, Darkelf172002, Davejohnsan, Davemcarlson, Decumanus, Delta 51, DerHexer, Dino, Dionysian35, Discospinster, Dougmcdonell, Dppowell, DriveMySol, Drmies, Dukeku, Dukemeiser, Edderso, Eddorn, Edward, Edwardsull, Egmontaz, El C, Emarsee, EncMstr, Epbr123, Explosionsnevermakeasound, FPAtl, Faustus37, Feydey, Fieldday-sunday, FloNight, Funnyhat, G. Capo, Galantra123, Galoubet, Grant Gussie, Grstain, Gscshoyru, Hayden120, HennessyC, Heqs, HiDrNick, Hide&Reason, Hike395, Hike796, Hmains, Hollyjanelle, Homely, Hooloovoo221, IRP, Ian Pitchford, Ice Cold Beer, Ilikewiki2, Indefatigable, Ipatrol, Iridescent, It Is Me Here, Itzcolotl, Ixfd64, J'onn J'onzz, J.delanoy, JCarriker, JDubman, JHeinonen, JStripes, JaGa, JackofOz, Jakewaage, Jakezing, JamesAM, Jao, Jengod, Jevansen, Jfollansbee, JimboV1, John.n-irl, JonHarder, Jonathanweber83, Joshuadfranklin, Jsayre64, JustAGal, Keilana, Keithonearth, Kellogg257, Kenneth Alan, Kevin Myers, Kevlar67, Klausness, KnowledgeOfSelf, Kopaka649, Koppenlady, Kross, Kroy the joker, Kudzu1, Kumioko, Kwamikagami, L3tt3rz, Lamb99, LastRemnant91, Latitude0116, LegolasGreenleaf, Leszek Jańczuk, Lightmouse, Linda Rider, Llywrch, Lockley, LogitechTHX, Lonewolf BC, Look2See1, Luftmann, MER-C, MONGO, MPF, Maelnuneb, Magioladitis, Mahanga, Makaristos, MarcoTolo, MarritzN, Marytrent, Matt Yeager, Matt0129, Maxee, MayaSimFan, Mduyns, Meelar, Megaman en m, Melchoir, Mikenorton, Miketherocker, Mild Bill Hiccup, Mimzy89, Minimac's Clone, Mintguy, Missoulian, Mo-Al, Mojoismog, Monterey Bay, Mooeydstupidcow, Moxy, Mr. Comodor, Myasuda, Mykej, NAW616, NE2, Nate Silva, Neo139, Neverquick, Nick Number, Night of the Big Wind Turbo, Nihiltres, Nimbusania, Niteowlneils, Noattention, NorCalHistory, Norcalal, NuclearWarfare, OhSoHeartless, Ok!, Opbhr, Osel, Ottre, Owen, Pearlwura, Penser,

PerdixPerdix, Peteforsyth, Pfly, Pharaoh of the Wizards, Philg88, Philip Trueman, PhnomPencil, Piano non troppo, Pinotgris, Plasma Twa 2, Plorimer, Pnw, Polylerus, Prodego, Provelt, Puget Sound, Quentin X, Qyd, R'n'B, Ravenna1961, Rawr, Rbrewer42, Reaper Eternal, RedWolf, RekishiEJ, Rising*From*Ashes, Rjensen, Rjferreiro, Robin Patterson, RockMFR, Romarin, Ronhjones, Rootrider, Roux, Rror, Ryan587, Ryandwayne, Ryuhaku, SJ Morg, SNIyer12, Salvio giuliano, Sandratosweetmcarter, SarahStierch, Sburke, SchfiftyThree, Schwnj, Scotchgit, Scriberius, Seatex, Seba5618, SebastianHelm, Sesu Prime, Sewing, Shannon1, Skew-t, Skookum1, Soliloquial, Soundersbloke, Sparrowhawk64, SpokaneWilly, Spoon!, Squash Racket, Steam5, SteveO, Steven Walling, Striker161, Sunsetpacific, Sunshine4921, Supergeekfreak, Superman181, Surachit, Symi81, TCY, Tassedethe, Teratornis, Tesscass, ThanatosXRS, The Interior, The Man in Question, The Thing That Should Not Be, The Ungovernable Force, TheTechieGeek63, Therealhazel, Theresa knott, Thomas Paine1776, Tim208, TimBentley, TimeClock871, Tmangray, Tom Morris, Tompw, Trezjr, Troublemaker400121, Trödel, Turtle55, UWDI ced, Udubwill, Ulric1313, UncleDouggie, Unless you, Usgnus, Uwieshu, VT hawkeye, Valfontis, Vanisaac, Velella, Vincentl, Volcanoguy, Vrenator, Wayward, Weedwhacker128, Weyes, WhoppeeSA, Wiarthurhu, Wiebe19, William Avery, Wombatcat, Woohookitty, X ile, YUL89YYZ, Ylem, Zapvet, Ziggy kid, Zutano, 800 anonymous edits

Confluence *Source*: http://en.wikipedia.org/w/index.php?title=Confluence *Contributors*: 16@r, Acs4b, Al.Neuland, Anders gorm, Andres, Anthony Appleyard, Apollo1758, Black Kite, Budelberger, Chanheigeorge, Cuaxdon, Cuvette, D6, Dan East, Danthemankhan, Dayg1110, Deeptrivia, Dekimasu, Doric Loon, Eleassar, ErikvanB, Euchiasmus, EugeneZelenko, Extermino, Fanatix, Filos96, Gdamasceno, Gerry D, Gilliam, Gits (Neo), Hlucho, Hmains, HolsteinPommern, Imnotminkus, Impreziv, Itai, JMS Old Al, JPMcGrath, Jake Wartenberg, Jkl, Jon Harald Søby, Jsbhavsar, Look2See1, Maniago, MarsRover, Matt Yeager, Mfranck, Mhockey, Michael Hardy, Mick Knapton, Mike Cline, Mikeblas, Mircea cs, Mkie, Mr3weebs, Muffuletta, Mukake, Mussapedia, Neelix, NeilDT, Nono64, Nopira, Paddad64, Pawl Kennedy, Peter Chastain, Pfly, PhnomPencil, Piotrus, Porphyro, Presidentman, Reynaldo en, Rsocol, Schaengel, Shawn in Montreal, Sherool, Smalljim, Steverwanda, That Guy, From That Show!, Transducción, Túrelio, Wiki-uk, Wikiuser100, Wilson44691, WinterSpw, ZRH, 54 anonymous edits

Willamette River *Source*: http://en.wikipedia.org/w/index.php?title=Willamette_River *Contributors*: Aboutmovies, AlexPlank, Amillar, Another Believer, Antandrus, Ben Ben, Bigjake, Bobblewik, Brad101, Brian Huffman, Bryan Derksen, BryceHarrington, CLW, Cacophony, Calmer Waters, Casliber, Cdc, Conversion script, Courcelles, DabMachine, Davidnwelton, Decumanus, EdChem, EncMstr, Esprqii, EurekaLott, Ezhiki, Finetooth, GermanJoe, Ghalleen, Glycerinester, Godfollower4ever, Godfrey Daniel, Good Olfactory, Gracenotes, Graham87, GreatOrangePumpkin, Grontesca, Gs 12, Gyrobo, Headbomb, Here.it.comes.again, Hike395, Hmains, Hrystiv, Hurricanehink, J.delanoy, Jimp, Jmchuff, JordanRL, Jsayre64, Justrob, Kkmd, Kmusser, Kvdveer, Kwamikagami, Lightmouse, Llywrch, Look2See1, Luk, MER-C, Malepheasant, Matt Gies, Max Schwarz, Mboverload, Motthoangwehuong, Mtsmallwood, Myasuda, Natpowning, Orygun, Peckvet55, Peregrine Fisher, Peteforsyth, Pfly, Philipnjones, R'n'B, RJHall, Rich Farmbrough, RobLa, Ruhrfisch, S. Neuman, SDY, ScottMainwaring, Shannon1, Sirlearnsalot, Skookum1, Squids and Chips, Stemonitis, Steven Walling, Teresaclin, Tesscass, Tide rolls, Titanium Dragon, Tsarevna, Tuspm, Ucucha, Valfontis, VerruckteDan, Vicki Rosenzweig, Vsmith, Wheelie Willy, Whhalbert, WikiDon, Will2020, Wnissen, Wsiegmund, Zondor, 97 anonymous edits

List of United States cities by population *Source*: http://en.wikipedia.org/w/index.php?title=List_of_United_States_cities_by_population *Contributors*: 08OceanBeach SD, 19994best, ACupOfCoffee, AJ Jennings, ALC Washington, AMP7, Aarenas, Acidulantes, Agceltics, AgnosticPreachersKid, Aivazovsky, Alakee, Alenag25, AlexR.L., Aliaksandrhn, Altenmann, Amateria1121, Andonic, Andy Marchbanks, Andyakamiztatwizta, Anoldtreeok, Anomalocaris, Antandrus, Antavius, Antony-22, Atacama, Atlantacitizen, Atler5264, Atlilmgaaustell, Av9, Avala, Averaver, Avgeiner, Awesomecleric, Axir, BD2412, Balanciert, Beland, Bellhalla, Berliner88, Bfkenney, BirgitteSB, Biruitorul, Bkonrad, Black Kite, Blainster, Blinkman44, Bmoredlj, Bobak, Bobby H. Heffley, Bobdud56, Bobo192, Bobster687, Bongwarrior, Bporter28, Bradleyandrews, BradlyRM, Brian the Editor, Brownings, BryanG, Brybry26, Buaidh, BuffaloChip97, Bullmoosebell, Bullshark44, Burns37, Burnwelk, Bus2Beezlebub, Butterboy, CIreland, CPacker, Calvin 1998, CambridgeBayWeather, CapPixel, Carlama025, Carpet9, Cdbravo, Celique, Charlesbell515, Charlesblack, Charlottedude, Charvex, Chinainfo2008, Chivafighter93, Chrishmt0423, Cincydude55, Circeus, Clalvarez14, Clev7, College Watch, Colonies Chris, Comayagua99, CommonsDelinker, Compaq1011, ComputerGuy, Conk 9, Cool Blue, Cordphaeton, Cougs2000, Coulraphobic123, Cpudude91, Crazymonkey1123, Criticalthinker, Ctjf83, Cuchullain, Cybercobra, D. Webb, D8010, Dailynetworks, Dale Arnett, Dallasborn&bred, Daniel Christensen, Daritto7117, Darth Mike, DeadEyeArrow, Denis C., Deor, Desomnambulist, Dhruvhemmady, DirkvdM, Dj1997, Djammurph, Djluci, Dkderick, Dkreisst, Dpm12, Dralwik, Dru of Id, Druid.raul, Drumguy8800, Dtcomposer, Duck25, Dude518, Dude976, Duece22, Eagle4000, EaglesFanInTampa, Eastlaw, Editor19841, Elephant200, ElizabethGStone, Elockid, ElockidAlternate, Eluchil404, Emeraldcityserendipity, Enorton, Epicadam, Esrever, EurekaLott, Evans1982, Everyoneandeveryone, Evhs00, Exsiiron, Extra999, FargomeD, Farras Octara, Fcb981, Flyingember, Fred1, Freddy Jade, Frietjes, Funandtrvl, GB fan, Gateman1997, GatorSlayerFSU, Gcarlisle, Gellersen, Geo0910, Geoman7773, GeorgHH, Georgia guy, Georgian, Ggskater, Giftlite, Giles Martin, Gimmetrow, Giorgi13, Golbez, Gooday.1, Govegan7, GraemeMcRae, Grayshi, Greatal386, GregFD3S, Grenavitar, Griffinofwales, Gymnos, Hammer1980, Hammersoft, Hawaiian732, Heegoop, Heimstern, Hequals2henry, Herostratus, Heyhey Hey, Hobbes747, Hoobershnooger, Huntster, Husond, I Wiki Edit, Iamtall47, Ibagli, IceUnshattered, Iceberg3k, Idaltu, Indy830, Indyguy, Infallibleflyingzombie, Iridescent, Itsandrewomg, J Milburn, J.delanoy, J117, JHarrelson, JHawk88, Jack Phoenix, Jajhill, Jalfredl, JamesBWatson, JamesMLane, Jaxonville, Jayron32, Jcrocker, Jd2718, Jdlddw, Jeff G., Jeffrey Gu, Jennavecia, JeremyA, Jericho735, Jerse, Jhendrix86, Jibbajabba, JimWae, Jimmy Slade, Jjron, Jleon, Jnorton7558, Jobroluver98, John K, Johnpacklambert, JonClee86, JonDE, Joshmacfie, Ju66l3r, Kaare, Kaldari, Kansan, Kar54, Karlos87, Katydidit, Kazzandra, Keilana, Kenneth Heal, KennyOdum3, Khatru2, King of Hearts, Klubbit, KnowitallWiki, Knuckles, Kristiansr, Kumioko, Kyem2010, Kylu, L Kensington, LabradorLover456, Lakercub, Lamrock, Leodmacleod, Levig12345, Lewis R, Liddlebigguy, LilMane, List of largest consumer markets, Lizy1090, Local hero, LonelyMarble, Loodog, Lord Pistachio, LtNOWIS, Luftmann, Luke D Taylor, LukeDT, Lyght, MJS04, MJS2300, MSClaudiu, Magicevan, Mariacer Cervantes, Marlinite, Martinez07, Master Frederique, Matt Kurz, Max Schwarz, MaxTraxx 82, Maximilian77, Mb93, Mboy1231, Mcattell, Mcbridelr, Meegs, Merchbow, Metroman2, Mgiganteus1, Michigan2101, Mikaey, Mikeaigotti8, Mildly Mad, Moka25, Moreau36, Mphifer254, MrPrada, Mrjpb104, Mwalcoff, ND58568, NailPuppy, Nakon, Ndrwatthedisco, Nickjtorres, Nishant12, Nkrosse, Nunocordeiro, Nyttend, Ohconfucius, Okanamondo, Okiefromokla, Orlady, Ottawa4ever, Owen, OxyMoronMinusOxy, ParaguaneroSwag, Pascal.Tesson, Paul August, PaulHanson, Paulinho28, Paxsimius, Peetlesnumber1, PeregrineAY, Peter E. James, Pgk, Philip Trueman, Phizzy, Phoenix2, Physicistjedi, PoccilScript, Polaron, Portalian, Postdlf, Ppntori, Prashanthns, Presidentman, Preuninger, Producercunningham, Protskugfjt98, Pumpkinart, Pwu2005, Quadell, Quentinisgod, Quiddity, RJN, Rackas321, Radon210, Random832, Rchilde, Rdrgz93, Redsfan5432, Redvers, ReignMan, Reinyday, Retropunk, RexNL, Rhallanger, Rich Farmbrough, Rlundi86, RockMFR, Roguegeek, Roif456, Ronwfree, RoyBoy, Rynoo, Ryoung122, Ryuhaku, Ryz05, SKIDS215, Sahmeditor, Samhuddy, Sampo Torgo, Sandwich Eater, Sawyerbiddle, Sayden, Scanlan, Schiptuin, Schwnj, Sepans, Sephiroth BCR, Sha721, Shakhan34, Shark96z, Shereth, Simpsondavid52, Skier Dude, Skinzfan23, Slo-mo, Soakologist, Someguy1221, Sonoran dweller, Soonerfever, South Bay, Southsanjose, Sparrowman980, Special-T, Spidey104, Spinach Monster, Spoofhound11, SpuriousQ, Starkirby64, Starlightmusic, StaticGull, Stealman, SteinbDJ, Stephanwehner, Steve44, Stevietheman, Stifle, Stlsportsfan2316, Stockton93, Strongbad1982, Suganthinjaffna, Sunderland06, Superman7515, Supername, Swarm, Tanthalas39, Taylor lover5, Tcncv, Telos, Tewfik, TexasAndroid, Than217, The Hut, The Mark of the Beast, The-Postman, TheHoosierState89, Thingg, Thomas Paine1776, Thomasalazar, Thomasmallen, ThunderbirdJP, TimBentley, Tjynyy, Tohd8BohaithuGh1, Tomlillis, Tompw, Tony1, Toonami Reactor, Tpbradbury, Transaspie, Travonmusic, Trevor MacInnis, Trorov, Turbokoala, Twistedlogic90, Tygrrr, UH Collegian, UbZaR, Uccs ur, Ufwuct, Ukexpat, Un sogno modesto, Uncia, Uninterested web user, UnitedStatesian, Unreal7, Unschool, Urbanative, Urbaneddie, Vchimpanzee, Veronique, Vicboy, Vlad788, Wars, Watkinsian, Wavehunter, Wenli, Wgbc2032, Whhalbert, WhisperToMe, Who, Wikidiculous, William milone, Williamevoss, Willisilly, Wjhonson, Wknight94, Woohookitty, Wpktsfs, XAtsukex, Xombie, XxTrillvillexX9, Yamamoto Ichiro, Yeng-Wang-Yeh, Zap261, Zebra848, Zers, Zinc2005, Zonafan39, Zzuuzz, Zzyzx11, , , 867 anonymous edits

British Columbia *Source*: http://en.wikipedia.org/w/index.php?title=British_Columbia *Contributors*: (jarbarf), -- April, -xfi-, 117Avenue, 200.191.188.xxx, 26 Ramadan, 382589454, 604555infofreak, A09A98986, Abootmoose, Academic Challenger, Accounting4Taste, Acroterion, Adashiel, Addshore, AdjustShift, Aenghus, Agamemnus, Agent 86, Agne27, Ahoerstemeier, Airplaneman, Alancookie, Alansohn, AlbertR, Alberth2, Aleenf1, AlexiusHoratius, Alksub, Alm93, Amaurus, Amazonien, AmericanIdolFreak22, Amikeco, Amorymeltzer, Andonic, Andrewlp1991, Andy Marchbanks, AndySimpson, Andycjp, Andyjsmith, Angela2109, Animum, Ann Stouter, Anna Lincoln, Antandrus, AntarcticPenguin, Anthonyd3ca, Antonio Lopez, Arctic.gnome, Aremith, ArielGold, ArmchairVexillologistDon, AshleyMorton, Atlan, AuburnPilot, Autospark, Avaragado, AwamerT, BCKanaka, Backspace, Bakabaka, Balcer, Balwinderdeep, BarretB, Barryob, Bartel1977, Batinde, Bcbuggy, Bearcat, BeardWand, Belligero, Berean Hunter, Betacommand, Betterusername, Big Adamsky, Big Bird, Billpeanut, Binksternet, Bkonrad, Blackswan12, BlindEagle, Bluerasberry, Bobanny, Bobblewik, Bobo192, Bogey97, BoomMan77, BoomerAB, Boosterdo, Bormalagurski, Bosonic dressing, Bowley88, BrainyBabe, Brandon5485, Brian Crawford, Brian0918, BrianGV, Brighterorange, Brion VIBBER, Brokendata, BrownHairedGirl, Bsadowski1, Buaidh, Buchanan-Hermit, BuickCenturyDriver, Burgundavia, Burn, Burnabyguy, Burningview, Bwilkins, Byroman14, C1T1Z3N3R4S3D, CALR, CDN99, CIS, CJCurrie, CJGB, CKlunck, Cahk, CalgaryWikifan, Caltas, CambridgeBayWeather, Camw, Can't sleep, clown will eat me, Canada1, CanadianLinuxUser, CanadianOne, Canuck85, Canuckle, Canusa44, Caponer, Captain panda, CardinalDan, Carewser, CaribDigita, Carlosfeaver, Carneylinguist, Catgut, Cchow2, Cherry1000, ChrisErb, Chun-hian, Circeus, Citicat, Civil Engineer III, Ckatz, Clam0p, Clayoquot, Cleduc, Clee7903, Cochranealtaguy, Colin dellow, CommonsDelinker, Contributor777, Converge, Conversion script, Cool Blue, Coralmizu, CordeliaNaismith, Core10117, Cornellrockey, Courcelles, CrazyDave, Crazyjoeda, Crc, Cremepuff222, Cs1maniac, Ctjf83, CutOffTies, DBaba, Dadude3320, Daniel, Daniel Collins, Daniel Quinlan, Danny, Darkcore, Darkildor, Darkwind6000, DaveDavis, Davewild, David Kernow, David Newton, Dbfirs, DeNoel, DeadEyeArrow, DeadeyeLefty, DeathM9, Decumanus, Deetdeet, Deflective, Dekimasu, Deleting Unnecessary Words, Delldot, Den fjättrade ankan, Denelson83, Denisarona, DerHexer, Deror avi, Desertsky85450, Dgrant, Dhp1080, Diberri, Dillard421, Dionysius321, DisambiguationGuy, Discospinster, Dlohcierekim, Dlohcierekim's sock, Dolovis, Donbert, Donutlover99, DoubleBlue, Doulos Christos, Download, Dreaded Walrus, Drhaggis, Droll, Dspradau, Ducksrox1213, DuncanHill, Dv82matt, Dúnadan, E Pluribus Anthony, EOZyo, EPICWHALE, ESkog, Earl Andrew, Eastpaw, Echinda, Eclecticology, EdBever, Eddie6705, Eeekster, Eekerz, El C, Electrobins, Emarsee, Emir34, Englishnerd, EnterorespAko, Entropy, Enviroboy, Epbr123, Epolk, Er Komandante, Eric-Wester, Erich Schmidt, Escape Orbit, Esprit15d, Essjay, Evb-wiki, EvelinaB, Evergreen 92, Everyone Dies In the End, Excirial, Eóin, FYW09, Facts707, Fail Fail Fail Fail, Fakhredinblog, Falcon8765, Fan-1967, FastLizard4, Favonian, Feliperinaldi, Fibonacci, FinalRapture, Finn-Zoltan, Fire 55, Firsfron, Fisherjs, Fishhead64, Floaterfluss, Flrn, FlyingPenguins, Foobaz, Formeruser-81, Formeruser-83, Formeruser0910, Fractions, Franamax, FrankCostanza, Frankjohnli, Fredb, Freshacconci, Friesguy, Frnortnr, Fumitol, Fusionmix, Fyyer, GHe, Gail, Gaius Cornelius, Galati, Galliforme, Galoubet, Gap9551, Garryoak, Gary King, GaryColemanFan, Gbutler77, Gcapp1959, GeoffreyAbubakary, Geoffunfun, George415, Geossepg, Giraffedata, GirasoleDE, Gjbloom, Gjs238, Glane23, GlenParkerHoldings2, Gmupps, Golbez, Goldfishbutt, Gonzob, Good Olfactory, GoodDay, Graham87, Grandeur0, GreatGreenArkleseizure, GreenGourd, Gregory.Jackson, Ground Zero, Gunnar Guðvarðarson, Gurch, Gurchzilla, Gwernol, HJ Mitchell, HalfShadow, Halthecomputer, Happy-melon, Harjk, Haunti, Hayden120, Headbomb, Hello32020, Highfields, Highvale, HistoryBA, Historyted, Hoolme, Hottentot, House1090, Howardchu, Howcheng, Hu Gadarn, Hugo999, Hussainsonic, Hvn0413, Hydrogen Iodide, II MusLiM HyBRiD II, IPharted, IW.HG, Ian Dunster, IceKarma, Ikh, IkonicDeath, Indefatigable, InverseHypercube, Iperson120, Iridescent, Island, Isthisthingworking, Ivirivi00, J.biro.beres, J.delanoy, JDCMAN, JDGODDARD, JHMM13, JJAlexander, JKZC16, JLaTondre, JYolkowski, Jackfork, Jackmingo, Jafarw, JamesBWatson, Jameshrynyshyn, JannisN, Janothird, Jao, Jauerback, Jauhienij, JavierMC, Jclemens, Jdobbin, Jennavecia, Jericho4.0, Jermdeeks, Jeronimo, Jewhater666, JillandJack, JimVC3, JimWae, JimboV1, JoanneB, Joe Schmedley, Joel Russ, Joeyconnick, John K, John wesley, Johnpacklambert, Jordan Elder, Joseph Solis in Australia, Josepp, Jossi, Jpbowen, Jplonka, Jrtayloriv, Jsalcedo,

Juanchito007, Jusdafax, Jusjih, KNHaw, Kafziel, Kaihsu, KamStak23, Kamran the Great, KanadaKid19, Karl-Henner, Katalaveno, Kayau, Kdau, Keefer4, Keep your fork, there's pie, Keilana, Keith Edkins, Kelvinc, Ken Gallager, Ken Wuschke, KenWalker, Kevin, Kevintoronto, Kgorman-ucb, Kilrogg, Kimiko, Kimmac, King Pickle, Kingturtle, Kintetsubuffalo, KirinX, Klicka833, Klingoncowboy4, Klparrot, Knight6130, Korandder, Kostmo, Koyaanis Qatsi, Kreditanstalt, Krich, Kross, Kseferovic, Ktwdallas, Kukini, Kungfuadam, Kurykh, Kwamikagami, Kyle1278, KyraVixen, L Kensington, LTD.Edition, Labordaychild, Lamanda, Landroo, Langherdup, Lastbetrayal, LeaveSleaves, Legalizeit, LegolasGreenleaf, Leighratcliff, Leszek Jańczuk, Levineps, Lexicon, Lightmouse, Lilpinoy 82, LindsayH, Little Mountain 5, LizardJr8, Lomn, Longhair, Lord Lugie, LordFenix, Loren.wilton, Ltmc, Luckyluke, LuigiManiac, Luigizanasi, Luisaallie, M.nelson, MJCdetroit, MONGO, MPF, MPerel, MacTire02, Maclean25, Macy, Madisonemery, Magister Mathematicae, Magog the Ogre, Makaristos, Mandarax, Maneater3033, Mannafredo, Marchartnett, MarcoTolo, Marek69, Martin451, Martinwguy, Masalai, Master shepherd, Materialscientist, Matt Yeager, Matthew Samuel Spurrell, MaxSem on AWB wheels, McCaster, Measure, Menchi, Michael B. Horner, Michael m8893, Midnight dreams123, Migospia, Mike Rosoft, Mike s, Mikeparkes76, Milessmart, Mindmatrix, Mirv, Mkdw, Mokwella, Montrealais, Moorespoint, Moreschi, Mountolive, Moxy, Mr Minchin, Mr.Badlands, Mrosscan, Mschel, Mtnphotographer, Mulp, Musse-kloge, Muzi, Myasuda, Myrtone86, Mzajac, NE2, NHRHS2010, NYKevin, Nakon, Nancy, Nat, Nath87, Nathaniel Christopher, Nations United, Nauticashades, NawlinWiki, Nehrams2020, Neilanderson, NellieBly, Nibuod, Nick Number, Nickyr2, Nikkimaria, Niteowlneils, No stems, no seeds, No such user, Noctibus, Noel Streatfield, NorseOdin, Northamerica1000, Notanoption, Npeters22, Nrusse, Nurmsook, Ogleby, OhSoHeartless, OhlundFan2, Ohthelameness, OldManRivers, Omnieiunium, Onebravemonkey, OrbitOne, Ottokarten, Outsidethewall, Owen, Oxymoron83, PKT, PMDrive1061, PS., PZFUN, Pacific Rim Vacation Rentals, Pak21, Patrick, Paul August, PaulHanson, Paulalexdij, Pb30, Pbryan, Pcb21, Pen-wu, Penmachine, Peregrine981, Peruvianllama, Peter I. Vardy, Petrb, Philip Trueman, Phraxis, Pieuvre, Pinoiako916, Piolinfax, Pmlineditor, Po' buster, PoliceChief, Polly, Poor Yorick, Poorpaddy, Prodego, Prolog, Protype2, Psmith, Puget Sound, Quadell, Quistnix, Qutezuce, Qwerty2562654, Qwerty6, Qxz, Qyd, R roy69, R'n'B, RA0808, RJaguar3, Rackstar, Rada, Ramaksoud2000, RandomCritic, Randydeluxe, Rashed, Raven in Orbit, Razorflame, RedWolf, Reginald Perrin, Regulov, Res2216firestar, RexNL, Rholton, Rich Farmbrough, Ricky-Roman, Rickymcsideburns, Ringan, Riotrocket8676, Ritual.entertainment, Rjensen, Rjwilmsi, Roarjo, RobTheGob, Robertgreer, Rockybiggs, Rockysmile11, Rogpyvbc, Roleypolinde, Romanskolduns, Ronnygosal, Ronz, Rory096, Rosiestep, Roux, RoyBoy, Royalguard11, Royboycrashfan, Rr parker, Rsrikanth05, RupertMillard, S h i v a (Visnu), S-Ranger, S0me l0ser, SD6-Agent, SJP, SMC, SNIyer12, Sam Korn, Sam Li, Sango123, Sarah 1940, Sasquatch, Savystooge, Saxifrage, Scanadiense, Schentler, SchfiftyThree, Schgooda, Schmidtmandaddy, Schumi555, Scorpion0422, ScottSteiner, Sdhonda, Sean William, Seaphoto, Seattle Skier, Senornumeroocho, Serapia4, Serte, Sewing, Shadowlynk, Shadyzminion123, Shanemcd, Shaners82, Sharrington, Shirik, Shoanoak, Sillyak, SimonP, Sinfulrevive, Sinfulsurvivor, Sinusoidal, Sionnach1, Sir Vicious, Sk wiki, Skeezix1000, Skizzik, Skookum1, Skraelinger, Skunkboy74, Sky Attacker, Slaja, Slakr, Slay01, Smsarmad, Snigbrook, Snowmanradio, Socially Isolated Hamster, Sonic99, Sonicwolf, Souky12, Soulresin, Spartaz, Specs112, SpookyMulder, SpuriousQ, SriMesh, Srice13, Staffelde, Staffwaterboy, StanC8, StatGraphs, Steam5, SteinbDJ, Steinsky, Stepkey77, Steve03Mills, SteveJothen, Storm Rider, StressedGasStationKid, Stroppolo, Stu pendousmat, Subtlescorpion, SunCreator, Sunray, Supasheep, SuperHamster, Supergeekfreak, Surachit, Susanrybar, Svens Welt, Szawski, Tabletop, Tagishsimon, Tai kit, Tawker, Taxee, TeaDrinker, Techman224, Ted Longstaffe, TedE, Telcourbanio, Teles, Template namespace initialisation script, Tesscass, The Interior, The Rambling Man, The Thing That Should Not Be, The Tom, The Transhumanist, Thebigjman, Themightyquill, Thingg, ThinkBlue, Thricecube, Throttle-junky, Tiddly Tom, Tide rolls, Tim1357, TimBray, Timc, Timkmak, Tmazhindu, Tobby72, Tomakiv, TotallyTempo, Tpahl, Tpbradbury, TransUtopian, Traxs7, Trevor MacInnis, Triage, TripleGemini, Tristan705, TrueRacer23, Tutmosis, Twiceuponatime, Twin Bird, Txr13, Tyb222, Tyronen, Ubergeekguy, Ucucha, Ufwuct, UncleDouggie, Unyoyega, Valfontis, Vancouverguy, Vanka5, Vanman2010, Vary, Vcelloho, Vclausse, Verrai, Versus22, Veryhip, Vicki Rosenzweig, Vidioman, Viriditas, Vishnava, Visik, Vmenkov, Voyager, Vranak, Vrenator, Vsalloti, Vsmith, Vzbs34, W. Frank, WFTD, Wackywace, Wafulz, Waggers, Wassupwestcoast, Wavelength, Wawaxi, Wayne Slam, Wenttomowameadow, Wesdub, Wetman, WhereAmI, Whytecypress, Wiikkiiwriter, WikHead, Wiki13, Wikiemon, Wikipagewatcher, Wikipediarules2221, Wikipelli, William Avery, Willking1979, Wimt, WojPob, Woodgascarsmagic, Woohookitty, Wulfnoth, Wzilker, X0fallonnn0x, XMan, Xxdriftwoodxx, Yamamoto Ichiro, Yamla, Yassineisgay321, Yath, YellowMonkey, Yidisheryid, Yintan, YubYub41, Z781, Zaphod Beeblebrox, Zhou Yu, Zoe, Zollerriia, Zorbzorb123, Zscout370, , 2002 anonymous edits

Portland metropolitan area *Source*: http://en.wikipedia.org/w/index.php?title=Portland_metropolitan_area *Contributors*: Aboutmovies, Acalamari, Acntx, Ajbenj, Anvilmedia, BD2412, Bobosheep24, Brian Huffman, Buddha24, Cacophony, Chris the speller, Clipdude, EME44, Eco84, Edwardsully, EncMstr, HangingCurve, Hike395, Hmains, Hrystiv, Jmchuff, Judge Trudy, Katt, KeithH, Kitch, Lampsalot, Leszek Jańczuk, Look2See1, Marky-Son, MojaveNC, Motthoangwehuong, NE2, Patriarca12, Peteforsyth, Poccil, Presidentman, Rich Farmbrough, RickK, Rjwilmsi, SJ Morg, ScottMainwaring, Secondarywaltz, Splanxna, SpokaneWilly, Srich32977, Steven Walling, Sumahoy, Taromsky, Tedder, Tesscass, Tygar, Valfontis, Woohookitty, 43 anonymous edits

Seattle *Source*: http://en.wikipedia.org/w/index.php?title=Seattle *Contributors*: -Majestic-, -js-, -xfi-, 0x6D667061, 1028, 123456789aaaaa, 198.144.192.xxx, 21655, 24.29.106.xxx, 2D, 2help, 6SJ7, 7, 777fortytwo, 8bitJake, A bit iffy, A. Parrot, A3RO, AEMoreira042281, AVand, Aalexander, Aaronbrick, Abendigoreebs, Acalamari, Achromatic, Acntx, Addshore, Admrboltz, Adrian.benko, Adsfasdfadf, Aep, Aflyhorse, Agne27, Ahoerstemeier, AirLiner, Aisha321, Ak282, Akcarver, Akers20, Alan Smithee, Alan.ca, Alansohn, Albanaco, Alex S, Alex1141, AlexThompson1991, Alexdeangelis86, Alexf, AlexiusHoratius, Alexseattle, Alkivar, Allmightypepsi07, Alphachimp, Altairisfar, Altenmann, Amakuru, Amazonien, Ambyent, Amcaja, Amchow78, AmericanFe, Americus55, Amstevens2006, Amulyabhatt23, Anclation, Andrejj, Andres, AndresTM, Andrevan, Andrew c, Andrew1718, Andrewlp1991, Android79, Andy Marchbanks, Anforowicz, Angela, AnhLe, Animum, Anjin-san, Annasw, Antandrus, Antoinefcb, Antonsb, Apollo1758, Appleboy, Appraiser, Arch dude, Architectsea, ArglebargleIV, Armsgravy, Arpingstone, Arrindel, Art LaPella, Artixxexes, Asatruer, Asdfg123, Aspects, Astropithicus, Astuishin, AtticusX, AuburnPilot, Auric, Avono, Awakeforever, AwesomoPhx, Ayrshire--77, Azza-bazoo, BAMP, BKP, Babooshbaboosh, Badbats, Bahnfrend, Bakke Graduate University, Bananaman134, Bandit495x, Bando26, BanyanTree, Baptisst, Barek, Barneca, Bartel1977, Barticus88, Bartledan, Baseball Bugs, Baseball319, Baylink, Baysidegirl and hugsarenice, Beenturns22, Beeshoney, Belka62, Belligero, Ben98112, BenFrantzDale, Bender235, Bensonee, Bevinbell, Bforte, Big Bird, Bigcheesebebbs, Bigfellahull, Bignd500, Billiam98, Billrnyc, Billywatson, Binksternet, Bissinger, Blackpride, Blaze33541, Blechnic, Bluespace, Bmclaughlin9, Boadrummer, Bob rulz, Bob2448, Bobak, Bobblehead, Bobblewik, Bobo192, Boisemedia, Boneillhawk, Bongwarrior, Boothy443, Bosnabosna, Bqmackintosh, Brad101, Bradfordschultze, Brady8, Brainboy109, Brainslug, Brandalone, Branden&papi, Brandon Mack, Brandon1978, Brennajones243, Brex2599, Brian0918, Brianhe, Brianneggk, Brighterorange, Brosq, Btm, Bubsty, Bumm13, Bunco man, Buphoff, Burntsauce, Butwhatdoiknow, C00w, CALR, CAVincent, CJLL Wright, CMLStillLovesDegus, CWenger, CZmarlin, Cacophony, Calegreen, CallMeAndrew, CallOfAndrew, Calvin 1998, Cam24, Can't sleep, clown will eat me, CanadianLinuxUser, CapitalSasha, Caprica, CardinalDan, Castor66, Cavrdg, Ccrraaiigg, Cdc, Ch'marr, CharlotteWebb, Cheesebob6000, Chipoliwog, Chr.K., Chris Roy, Chris the speller, Chrishelms, Christy747, Chubbles, Chuckwatson, Cianpower, Cjs56, Ckatz, Cliffb, Closedmouth, Cody5882300, Collegepres, Colonies Chris, Colors99, CommonsDelinker, Compdude123, ComputerGuy, Confiteordeo, Connor1995 02, Conti, Conversion script, Cool Stuff Is Cool, Cool-guy357, Coolmarcell, Cordell, Corvus cornix, CougMichelle, CoulterTM, Coz 11, Crabsoup, Craignielsen, CrazyC83, Crd721, Cruiser1, Cubicle BOB, Cutlass Supreme, Cxz111, Cyanidethistles, Cyrus XIII, DASA2, DAZNking, DVD R W, Dabomb87, Daedalus-Prime, Dale Arnett, Daltanimal, Daltnpapi4u, Dame Agatha, Danfan5, Daniel J. Leivick, Danieyek, Dansmitty2, Dark Shikari, Dark Tea, Dark Tichondrias, Darkohead, Darth Panda, Dasani, David Foster, David Newton, DavidLevinson, Davidlawrence, Davidwr, Dbfirs, Dbriski13, Ddgonzal, Ddnile, Deacon of Pndapetzim, DeadEyeArrow, Deanos, Debresser, Delirium, Delius1967, Demong, Den fjättrade ankan, Dennis7410, Der Rote Hahn, DerBorg, DerHexer, DerRichter, Deramisan, Derek.cashman, DesertAngel, Despeaux, Dhamil07, Dharmabum420, DianaRuiz, Didaktron, Diderot's dreams, Digitallib, Dikaiopolis, DinosgoRAWR, Dinoveronica, Dionysian35, Discospinster, Dmodlin71, DoYourBusiness, DocWatson42, Dockcharlotte, Docu, Dolphin Jedi, Dopeheadonmoped, Dori, DoubleBlue, Doulos Christos, Download, Downtown dan seattle, Downtown005, Dpa77, DrKeirnan, DrKiernan, DragonflySixtyseven, Draugen, Dreadstar, Drmies, Drumguy8800, Dryazan, Dschwen, Dsnow75, DubbaG, Dudesleeper, Duja, Dungodung, Dustfinger1fan, Dustingc, Dycedarg, EOBeav, ERcheck, EWS23, Eadgce, Eaglesmarty08, Eastlaw, Eco84, Ed g2s, Editrrr, Edward, Eellestad, Eequor, Eggtopolis, Ehccheehcche, Einstein cs, Ekmtap, El C, ElKevbo, Eldang, Emeraldcityserendipity, Empty Buffer, EncMstr, Enfusia, Englishwildman, EoGuy, Epbr123, Epicadam, Eptin, Equinox17, Eric-Wester, Ericd909, Erik4, Erik9, Erikrem, Eurosong, Evandro Davis, Everyking, Everynameisalreadytaken10001, Evil Monkey, Evnmrrs1, Excel23, Exir Kamalabadi, Extransit, Ezeu, F, F15x28, Facts707, Falcon8765, Fannybandits, Fastestdogever, Faz90, Femto, Feydey, FictionalCharacterGuy, Fifthhorizon, Finlay McWalter, Fipher, Firsfron, Flata, Fleminra, Florian Blaschke, Floridagators, Fluffernutter, ForrestCroce, Forthac4, Fostermarkd, Frankie0607, Freakofnurture, Fred Bradstadt, Funandtrvl, Funktasm, Funnyhat, Furrykef, Fvasconcellos, Fyrael, GDibyendu, GManNickG, GNRY09, GS121389, GSTQ, GTBacchus, Gaff, Gail, Gaius Cornelius, Galoubet, Gamerx456, Garrepi, Garrett Albright, GastonBruna, GatorFTL, Gatorman, Gene Nygaard, GeneralCheese, Genevievemonty, Genyaz, Georgewilliamherbert, Georgia guy, Geosultan4, Gepaulsen, Gerry D, GetYooRight, Ghazali01, Ghwatts, GiZiBoNG, Gil Gamesh, Gilliam, Ginsengbomb, Gjking, Glacier109, GoDot, GodKronos1337, Godlord2, Gogo Dodo, Golbez, Goldom, GoneAwayNowAndRetired, Goodnightmush, Goodsteve00, Gotonash, Gr8opinionater, Gracenotes, GraemeL, Graemelc, GrafZahl, Graham87, Granburyian, Grant65, Greekafella, Greenshed, Gremo32, Grey Wanderer, Greyyhawkk, Grimmfire, Grseattle, Gscshoyru, Guesty1, Gurch, Gusbenz, Gwernol, Gz33, HUIMARK, Hadal, Hamiltondaniel, Hammey12345, HandsomeFella, HappyLogolover2011, Harland1, Haruo, Haseo9999, Hateless, Hayabusa future, Hdt83, HelloWorld206, Hellosandimas, Henry Flower, HereToHelp, Hersfold, Hetar, Hi878, Hike395, HillValleyTelegraph, Hinto, Hippo master666, Hirudo, Hmains, Hmbdude13, Hmcnally, Hmoul, HockeyLovingHoser, Hockeynut11, Hojimachong, Honeyhucket, Hooriaj, Hoppercool, Hoshinotzora, Howardw, Howcheng, Hubertfarnsworth, Huhwaitwhoareyou, Hunter Kahn, I'M TEH REAL NUMBER ONE, ICAPTCHA, Ian Morris, Iknow12345, Illspirit, Imbrainfever, Imlno, Imnotti, Independent Journalist, Indopug, InfoProvider, Infoeco, Infrogmation, Isaacada1, Isatemple, Ixfd64, J.delanoy, JDubman, JHawk88, JPD, JPMcGrath, JaGa, JabberWok, Jackofh3arts, Jaeden24, Jakeroot, Jameslarousse, Jamestl, Jamies, JasonAQuest, Jazzkittykat, Jblazz21, Jbuw, Jclemens, Jeancey, Jeepday, Jeff G., Jeffmanson, JeffreyAllen1975, Jelson25, Jengod, Jennavecia, Jerrod1, Jesse-inside, JesseW, Jesseone, Jesser07, Jessicavoelker, Jfknrh, Jfollansbee, Jhendin, Jiang, JimWae, Jimmey, Jimmy Slade, JimmyCrackedCorn Makes A Complaint, Jimp, Jiy, Jkaplan, Jkiang, Jkl sem, Jleon, Jlhiowa, Jllm06, Jmabel, Jmaddux21, Jmlk17, Jo Weber, JoanneB, Jobe9, JocobFTB2, Joelr31, Johaen, John, JohnInDC, Johnny Rockfort, Jok2000, Jon314, Jonarthur, Jonchapple, Jonel, Jonemerson, Jons63, Joowwww, JordoCo, Jordon Kalilich, Josef Pampalk, Josephabradshaw, Joshua Scott, Joshuadfranklin, Joshulove, Jossi, Joyous!, Jpers36, Jplatts, Jpm468, Jrauser, Jsayre64, Jstein16, Jtmorgan, Ju66l3r, Judahdavis, Jugbug2, Juliancolton, Jusdafax, Jutras90, Jwinters, Jwrosenzweig, K.lee, KGasso, KLIMATT, Ka Faraq Gatri, Kaiser matias, Kaiwhakahaere, Karlwiegand, Karol Langner, Kastanday, Kateshortforbob, Katydidit, Kazamm, Kbdank71, Keegan, KeithTyler, Ken Wuschke, Kerotan, Kesla, Kevin Rector, Kf4bdy, Kindall, King 4rthur, King of Hearts, Kingpin13, Kingturtle, Kirach98, KirbyOlson, Kkmd, Klhill1962, KnowledgeOfSelf, Knut17, Kogoro Mouri, Koolkatiekat, Koppenlady, Kristen Eriksen, Kristiansr, Kristof vt, Ktr101, Kukini, Kukuman, Kungfuadam, Kuru, Kurykh, Kuyabribri, Kwamikagami, Kyoko, Köln, L Glidewell, L Kensington, L3tt3rz, LLarson, LQ, Lacrimosus, Lajsikonik, LakeFP, Laonoodlekeemow, Larmitstead, LarryGilbert, Latinboy1976, Laurenrose3091, LeaveSleaves, Lecollye, Leevanjackson, Legoktm, LegolasGreenleaf, Les woodland, Leujohn, Levineps, Lewastate, Lfh, Lifeofbrian, Lightmouse, LindsayH, Liquid Hawk, Litheapo, Little Mountain 5, Llywrch, Lmessenger, Local hero, Lol1212, Lolacirill, Lolacirillo3, LonelyMarble, Loodog, Look2See1, Loopin123, Loopknow, Lotje, Lowbart, Lquilter, Ludvigant, Luk, Lukobe, Luna Santin, Lupin, Lupinelawyer, Lupo, Lurlock, MC10, MCWoody65, MER-C, MJCdetroit, MK8, MLRoach, MPerel, Mactabbed, Mad Max, Madchester, Madeingermany, Magioladitis, Magister Mathematicae, Magus018, Malt Liquor, Manticore, Marc-Olivier Pagé, Mareino, Marek69, Margareta, Marinersfanbart, Mark, Mark83, MarkGallagher, MarkHab, Markandbarbara, Marklevywcf, MarmadukePercy, Maroux, MarritzN, Martarius, Martin451, MartinDK, Martyn Lovell, Masaruemoto, Masonfree40, MastCell, Mastermorfeu, Materialscientist, Mathpianist93, Matiasp, Matrix99, Matt Yeager, Matt3589, MattWright, Mattisse, Mav, Mbinebri, McNeight, McSly, Mccullough74, Mcmax77, Mddake, Mdkarazim, Meanjoe, Measure, Meepster, MegX, Megan1967, Mehdig, Mel Etitis, Mendaliv, Meteorologists, Mh29255, Mhbourne, Michael Devore, Michael Snow, Michael93555, Michak, Middayexpress, Midwesterndynasty, Mightymights,

Mikaka, Mike Halterman, Mike Selinker, Mikealrogers, Mikeewen 19, Mikenorton, Mikewick, Millbrooky, Millionare101, Minimac's Clone, Minna Sora no Shita, MisfitToys, Misterwhite, Mkdw, Mlenoirh, Mlouns, Mnicolosi, Modernmonalisa, MojaveNC, Mojoworker, Moncrief, Monegasque, Monstermudder78, Monterey Bay, Mooch14, Moreau36, Morganfitzp, Morwen, Moscatanix, Mozy8, Mpalmer22, Mpj17, MrAMMOHEAD, Mrarfarf, Mroliver83, Msweeny, MuZemike, Muchness, Muggleflucka, Mujilakama, Mulad, Mwanner, Myasuda, N5iln, N734LQ, NE2, NGC6254, NLuchs, NRS11, Naddy, Nadroj1997, Nakon, Nankai, Nards911, Nasmformyzombie, Natalie Erin, Natalya, Nate Silva, Nathanmakan, NavalPower, Neal Finne, Necronaut, Nedolson, Neiboeneri, Nekohakase, Netjeff, Neutrality, Nev1, NewEnglandYankee, Newlove oldfaith, Newnoise, Ng.j, NickBush24, Nightalive, Nightkey, Niki98144, Nilmar2, Nisetpdajsankha, Niteowlneils, Nnivanov, Nolenjb, Noren, North Madison, Nothingofwater, Nova77, Novis-M, Npalleroni10, Nrswanson, Nsimpson1509green, Nuberger13, NuclearWarfare, Nzpcmad, O% interest, OMHalck, OOODDD, Obersachse, Obgydd, Ocee, OhSoHeartless, Ohconfucius, Ohnoitsjamie, Okiefromokla, Old Guard, Olivier, Omicronpersei8, Ommnomnomgulp, OnePt618, Opie, Orestek, Oro2, Orthographer, Oska, Ottavina, Owen, Ows2011, Oxymoron83, PDH, PMDrive1061, PSzalapski, Paeris, Pak21, Pasadena91, Pastoredwards, Pat iz koool, Patleahy, Patrick, Patricknoddy, Paul August, Paul Richter, Pazpark, Pcbene, Pdxfulham, Peanutbutterjellytime4, Pebbens, Pen1234567, Pengortm, Penser, Pentawing, PeregrineAY, Perfect Tommy, Persian Poet Gal, Peter Isotalo, Peter135, PeterSymonds, Pfly, Pgan002, Pharos, Phe, Philip Trueman, PigFlu Oink, Pigboats, Pinethicket, Piotrus, Piuro, Pokeyman1, Polaron, Ponydepression, Pookster11, Popcultureguy, Popebrak, Poppalnu, Poppapo, Possum, Post-it-88, Postdlf, Postoak, Potaaatos, Ppak, Priestdi, Prolog, Promethus6, Proofreader77, PseudoOne, Psy guy, Pumpkincrook, Pxma, Qbmessiah, Quadell, Quadratum, QuarterSpanishRevolutionist, Quasipalm, Qxz, RCPayne, RFD, RJN, RWallace1776, RXPhd, Raccoon Fox, Rachelkarin, RadicalBender, Radicalsubversiv, RaggieSoft, Railer 127, Raime, Rainwater22, Ram-Man, RandySavageFTW, Rattlhed, Raul654, Rawbeedee, RaySys, Rdicker, Rdsmith4, Reach Out to the Truth, Realestateiscool, Rebecca, Red Baboon, Redwolf24, Regardless143, Regulate58, Remember the dot, RenoTC11, Reptilizer, Retired username, Revcph1958, Rewinn, RexNL, Reyellogreebl, Reywas92, Rhatsa26X, Rich Farmbrough, Richardaedwards, Richiesullivan, Richo1007, Rick Block, Rileyrj, Rjd0060, Rjensen, Rjwilmsi, Rkt2312, Rlfb, Rlogan2, RobLa, Robchurch, Robdurbar, Robert Skyhawk, RobertG, Roberta F., Robertb-dc, Robin Patterson, Rockfang, Rockstar915, Roke, Ronabop, Ronz, Rosestiles, RossPatterson, Rossdegenstein, RossweisseSTL, Roxi2, Royalguard11, Rrajchel, Rstevec, Rularue, RussellMcKenzie, Rwerner, Ryulong, S, S.hammarlund, SDC, SPUI, ST47, SWAdair, Safesler, Sakkura, Sam Francis, Sam42, Samba6566, Samuelsen, Sandy60, SandyGeorgia, Sandylee253, Sanfranman59, SarekOfVulcan, Sasbens, Savidan, Sbscavs8, Scarequotes, SchfiftyThree, SchmuckyTheCat, Schnell, SchuminWeb, Schzmo, Scottblanco, Scriptwriter, Seahorseruler, Sean WI, Sean mc sean, Searchwriter, Seasound12, Seatex, Seattle Daily Photo, Seattle's, Seattle2008, Seattle99, SeattleArch, SeattleChronic, SeattleThyme, SebastianHelm, Sedna1000, Sedna10387, SelfStudyBuddy, Severbeck, Sfdan, Sfmontyo, Sgodth, Shadowjams, Shane Lin, Shanes, Shawnsawb, Shell Kinney, Sherrold, Sherwelthlangley, Shimgray, Shinmawa, Shoaler, Shoopza, Shortride, Shreshth91, Shyam, Sicamous, Silverback, SimonArlott, SimonP, Simulcra, SiobhanHansa, Sir Richardson, Siradia, Skizzik, Skyman123, Slakr, Slayemin, Slightsmile, Slimey.limey, Slugojake, Smallbones, Smith03, Smlowe5, Smrose, SnowConeYellow, Snowolf, SoCalSuperEagle, Socially Isolated Hamster, Sonicology, South Bay, SpaceyDave, SpamMan, SpikeTorontoRCP, SquidSK, Srich32977, Steggall, Steinsky, Stephenwood, Steve03Mills, Steven Zhang, StevenBradford, StevensCourt, Strebe, Strongbad1982, Student7, Stwalkerster, Suffusion of Yellow, Sumsum2010, Sundodger, Sunray, Super Martyo Brother, Superman181, Supermandan, Supermarioz1, Supershippy, Surachit, Swedish fusilier, SweetCarmen, Symi81, TEG24601, TUF-KAT, Taifarious1, Tall copy, Tartessos75, Taurion, Tawker, Tbhotch, Tboy206, Tcncv, TechBear, TedTheUnique1, Tedder, Tellyaddict, Tenebrae, TerminalPreppie, Terriblety, Tesscass, TexasAndroid, Texture, The Interior, The Man in Question, The Thing That Should Not Be, The Ungovernable Force, The Universe Is Cool, TheFreeloader, TheLeopard, Theda, Thesetrixaintforkids, Theshibboleth, Thewikigod, Thingg, Thomas Paine1776, Thomasdelbert, Thomasedavis, Thoreaubred, Tibetan Prayer, Tide rolls, Tifflea, Tillicum's finest!, Tim Ivorson, Timothylord, Tiroj1, Titoxd, Tj11hd19, Tobby72, Tomakiv, Tommy2010, Tomos, Tonster, Tony1, Tonyle, Tooto, Topsydog, Tothebarricades.tk, Tpbradbury, Tradnor, Trafton, Tralalalalala, TransUtopian, Traveling Man, Travis.Thurston, Trialsanderrors, Trodaikid1983, Trojan51, Trrill, Tsg946, Ttownfeen, Tudorjason, Tukwilan, Tuxmelvin, TwoKatz, UWDI ced, Uagehry456, Udubwill, Ukabia, Un sogno modesto, Uncle Dick, Unixan, Unschool, Uofmfan909, Uris, Useight, User2004, VMS Mosaic, Vald, Van helsing, Vcelloho, Vclipper, Veinor, Ventur, Ventusa, Veronique, VgerNeedsTheInfo, Vh20, Viajero, Vicarious, Vicki Rosenzweig, VinnieSayzEeetMe, Violetriga, Viper5hdz, Vranak, Vsmith, W.stanovsky, WAvegetarian, WJBscribe, WWGB, Waggers, Wakx, Walker9010, Walkerma, Wall.dylan, Waqcku, Wavelength, Wayne Slam, Wayward, Westernyankee, Westonmr, Wfgiuliano, Whalewriter, WhereTimeStandsStill, WhisperToMe, White 720, White Trillium, Who, Wibach, Wiki alf, Wikibarista, Wikieditor06, Wikihipster, WildlifeAnalysis, Wildthing61476, Will Beback, William Allen Simpson, Willingandable, Windowsvista32, Wittyname, WizardDuck, Wknight94, Woodelfking, Woohookitty, WookieInHeat, Work permit, Worm That Turned, Wsiegmund, Wsupress, Wtmitchell, X-Weinzar, XLerate, Xavierschmit, Xdenizen, Xiaoyu of Yuxi, Xsarahrahx, Xyzzyplugh, Xyzzyva, YPLeroux, Yakiv Guck, Yalbik, Yayme9, Ycdkwm, Yeknom der, Yngvadottir, Yourstruly598, Yuckfoo, Yworo, Z1nemo, Zachattack088, Zanzibar12345654321, Zecyor, ZephyrKylin, Zerocard13, Zidonuke, Ziggurism, Zoe, Zundark, Zutano, Δ, Александър, 3161 anonymous edits

Cedar Mill, Oregon *Source*: http://en.wikipedia.org/w/index.php?title=Cedar_Mill%2C_Oregon *Contributors*: Aboutmovies, Bob Palin, Bumm13, Carlossuarez46, Cdrmil, Charles Matthews, DanielCristofani, Disneywizard, Kcordina, Lightmouse, Llywrch, MPF, Motthoangwehuong, NE2, Nyttend, Pearle, Ram-Man, SJ Morg, Syrthiss, Trafton, User2004, Valfontis, Virginia Bruce, 15 anonymous edits

Washington County, Oregon *Source*: http://en.wikipedia.org/w/index.php?title=Washington_County%2C_Oregon *Contributors*: Aboutmovies, Acalamari, Acntx, Alibguy, Amikake3, AshyLarry, Backspace, Bobblewik, Buddha24, CanisRufus, Cfrjlr, Colonies Chris, CommonsDelinker, Conversion script, D6, Davidt100, Dawnseeker2000, Dscott1349, Emeraldcityserendipity, EncMstr, Evil saltine, Ex-LasVegas-Girl, Fitzaubrey, Funnyhat, Glacier109, Graham87, Hobo Dave, Hyperfast, Ipoellet, Justrob, Leslie Mateus, Libereco Li, Lightmouse, Lincolnite, Llywrch, Millicanmd, Minesweeper, Mlepisto, Monegasque, Motthoangwehuong, NE2, Northwesterner1, Numbo3, Nyttend, Oat Lock, OregonD00d, Owen, Pablothegreat85, Peteforsyth, Portlandvisitorguide, Postdlf, PseudoOne, PurpleSunflower, Quidam65, Ram-Man, Rettetast, Rhatsa26X, Rhiannon-wcva, Rich Farmbrough, Rjwilmsi, Rob Salzman, Rootbeer, Rossdegenstein, SJ Morg, SPUI, ScottMainwaring, Sgt Pinback, SirEditALot, Smallbones, Smickie765, TUF-KAT, Tabletop, Template namespace initialisation script, Theiceberg, Tim!, TimeClock871, Valfontis, Whhalbert, Worldenc, Zinc2005, 53 anonymous edits

Lake Oswego, Oregon *Source*: http://en.wikipedia.org/w/index.php?title=Lake_Oswego%2C_Oregon *Contributors*: Aaronbrick, Aboutmovies, Acntx, Ahalley, Ajbenj, Alansohn, Alexbbard, Andre Engels, Anna Lincoln, Arthurape, Ben D., BigFatBuddha, Bkonrad, Blamblamblam, Blankshot218, Bobblewik, Bobo192, BryceHarrington, Bumm13, Californiadrew, Cat Food, Cg41386, Dale Arnett, Darloftus, Dbc12, Deflective, Emeraldcityserendipity, EncMstr, Erikmcadams, Esprqii, ExpLo7, Fabrictramp, Finetooth, Frank101, Fæ, Gerry D, Gregoe86, Harland1, Harpchad, Henryodell, Hmains, HumbleGod, Hushpuckena, IBook of the Revolution, Jackal4, JeLuF, Jllm06, Jmchuff, Jnk, Jrowh, Kotra, Kwamikagami, Laker1man, Lolaker22, Luvv1234, MJCdetroit, MJHankel, MPerel, Mithridates, Moncubus, Motthoangwehuong, Mr Rogers, NE2, NHRHS2010, No1lakersfan, Northamerica1000, Nyttend, Oscarthecat, Owen, PDXblazers, Parhamr, Pdcook, Peteforsyth, Plazak, Quuxplusone, RJFJR, Ram-Man, Rich Farmbrough, Rror, SJ Morg, Sammanilla, ScottMainwaring, Scoty6776, Seth Ilys, Shenme, SimonP, Sksawin, Sky Attacker, Spellmaster, Stepheng3, Stevejohnson3215468, Syrthiss, Tassedethe, Tdowling, Tedder, Template namespace initialisation script, Tesint, Tesscass, Tfeinman, The Thing That Should Not Be, Thebarkitect, Tintin88, Trafton, TruthseekerB44, Ulmanor, Valfontis, VinceBitong, Voyagerfan5761, Wasted Time R, WhisperToMe, Wimt, Yamamoto Ichiro, 203 anonymous edits

Image Sources, Licenses and Contributors

File:Portland and Mt Hood.jpg *Source*: http://en.wikipedia.org/w/index.php?title=File:Portland_and_Mt_Hood.jpg *License*: unknown *Contributors*: User:Amateria1121

File:Flag of Portland, Oregon.svg *Source*: http://en.wikipedia.org/w/index.php?title=File:Flag_of_Portland,_Oregon.svg *License*: unknown *Contributors*: Mysid

File:Seal of Portland OR.png *Source*: http://en.wikipedia.org/w/index.php?title=File:Seal_of_Portland_OR.png *License*: unknown *Contributors*: Bestiasonica, Coldpas, Ed veg, Kilom691, Little Savage

File:Multnomah_County_Oregon_Incorporated_and_Unincorporated_areas_Portland_Highlighted.svg *Source*: http://en.wikipedia.org/w/index.php?title=File:Multnomah_County_Oregon_Incorporated_and_Unincorporated_areas_Portland_Highlighted.svg *License*: unknown *Contributors*: Arkyan

file:Usa edcp location map.svg *Source*: http://en.wikipedia.org/w/index.php?title=File:Usa_edcp_location_map.svg *License*: unknown *Contributors*: User:Uwe Dedering

File:Red pog.svg *Source*: http://en.wikipedia.org/w/index.php?title=File:Red_pog.svg *License*: unknown *Contributors*: Anomie

File:Portland Oregon in 1890.jpg *Source*: http://en.wikipedia.org/w/index.php?title=File:Portland_Oregon_in_1890.jpg *License*: unknown *Contributors*: U.S. Department of Defense

File:WillametteRvrPano edit.jpg *Source*: http://en.wikipedia.org/w/index.php?title=File:WillametteRvrPano_edit.jpg *License*: unknown *Contributors*: User:Cacophony

File:Magnify-clip.png *Source*: http://en.wikipedia.org/w/index.php?title=File:Magnify-clip.png *License*: unknown *Contributors*: User:Erasoft24

File:HawthorneBridge-Pano.jpg *Source*: http://en.wikipedia.org/w/index.php?title=File:HawthorneBridge-Pano.jpg *License*: unknown *Contributors*: User:Cacophony

File:Portland_Night_panorama.jpg *Source*: http://en.wikipedia.org/w/index.php?title=File:Portland_Night_panorama.jpg *License*: unknown *Contributors*: User:Fcb981

File:portland.png *Source*: http://en.wikipedia.org/w/index.php?title=File:Portland.png *License*: unknown *Contributors*: User:Sean Kelly

File:IntnlRoseTestGarden.jpg *Source*: http://en.wikipedia.org/w/index.php?title=File:IntnlRoseTestGarden.jpg *License*: unknown *Contributors*: User:Cacophony

File:TomMcCallWaterfrontPark.jpg *Source*: http://en.wikipedia.org/w/index.php?title=File:TomMcCallWaterfrontPark.jpg *License*: unknown *Contributors*: User:Cacophony

File:SchnitzerAtNightFront.jpg *Source*: http://en.wikipedia.org/w/index.php?title=File:SchnitzerAtNightFront.jpg *License*: unknown *Contributors*: User:Cacophony

File:Portland Oregon - White Stag sign.jpg *Source*: http://en.wikipedia.org/w/index.php?title=File:Portland_Oregon_-_White_Stag_sign.jpg *License*: unknown *Contributors*: User:Steve Morgan

File:Hawthorne st.jpg *Source*: http://en.wikipedia.org/w/index.php?title=File:Hawthorne_st.jpg *License*: unknown *Contributors*: Original uploader was Kyle Burris at en.wikipedia

File:StumptownCoffeeDivision.jpg *Source*: http://en.wikipedia.org/w/index.php?title=File:StumptownCoffeeDivision.jpg *License*: unknown *Contributors*: User:Cacophony

File:RoseGardenArenaS.jpg *Source*: http://en.wikipedia.org/w/index.php?title=File:RoseGardenArenaS.jpg *License*: unknown *Contributors*: User:Fcb981

File:Portland house price index.gif *Source*: http://en.wikipedia.org/w/index.php?title=File:Portland_house_price_index.gif *License*: unknown *Contributors*: Amy Vander Vliet

File:UnionStationPortland.jpg *Source*: http://en.wikipedia.org/w/index.php?title=File:UnionStationPortland.jpg *License*: unknown *Contributors*: User:Cacophony

Image:PortlandTriMetMAX.jpg *Source*: http://en.wikipedia.org/w/index.php?title=File:PortlandTriMetMAX.jpg *License*: unknown *Contributors*: User:Alphalife

Image:PortlandStreetcar5.jpg *Source*: http://en.wikipedia.org/w/index.php?title=File:PortlandStreetcar5.jpg *License*: unknown *Contributors*: User:Cacophony

Image:PortlandTramCar3.jpg *Source*: http://en.wikipedia.org/w/index.php?title=File:PortlandTramCar3.jpg *License*: unknown *Contributors*: User:Cacophony

File:PortlandCityHall.jpg *Source*: http://en.wikipedia.org/w/index.php?title=File:PortlandCityHall.jpg *License*: unknown *Contributors*: User:Cacophony

File:PortlandOR-aerial.jpg *Source*: http://en.wikipedia.org/w/index.php?title=File:PortlandOR-aerial.jpg *License*: unknown *Contributors*: Edward, Mattes, Werewombat, Xnatedawgx

File:PDX1966PGEplant.jpg *Source*: http://en.wikipedia.org/w/index.php?title=File:PDX1966PGEplant.jpg *License*: unknown *Contributors*: Original uploader was Rw rynerson at en.wikipedia. Later version(s) were uploaded by Parhamr at en.wikipedia.

File:Portland population growth.png *Source*: http://en.wikipedia.org/w/index.php?title=File:Portland_population_growth.png *License*: unknown *Contributors*: User:Jsayre64

File:MarquamBridgeHill.jpg *Source*: http://en.wikipedia.org/w/index.php?title=File:MarquamBridgeHill.jpg *License*: unknown *Contributors*: User:Cacophony

File:Flag of Israel.svg *Source*: http://en.wikipedia.org/w/index.php?title=File:Flag_of_Israel.svg *License*: unknown *Contributors*: AnonMoos, Bastique, Bobika, Brown spite, Captain Zizi, Cerveaugenie, Drork, Etams, Fred J, Fry1989, Geagea, Himasaram, Homo lupus, Humus sapiens, Klemen Kocjancic, Kookaburra, Luispihormiguero, Madden, Neq00, NielsF, Nightstallion, Oren neu dag, Patstuart, PeeJay2K3, Pumbaa80, Ramiy, Reisio, Rodejong, SKopp, Sceptic, SomeDudeWithAUserName, Technion, Typhix, Valentinian, Yellow up, Zscout370, 31 anonymous edits

File:Flag of the Republic of China.svg *Source*: http://en.wikipedia.org/w/index.php?title=File:Flag_of_the_Republic_of_China.svg *License*: unknown *Contributors*: 555, Bestalex, Bigmorr, Denelson83, Ed veg, Gzdavidwong, Herbythyme, Isletakee, Kakoui, Kallerna, Kibinsky, Mattes, Mizunoryu, Neq00, Nickpo, Nightstallion, Odder, Pymouss, R.O.C, Reisio, Reuvenk, Rkt2312, Rocket000, Runningfridgesrule, Samwingkit, Sasha Krotov, Shizhao, Tabasco, Vzb83, Wrightbus, ZooFari, Zscout370, 73 anonymous edits

File:Flag of South Korea.svg *Source*: http://en.wikipedia.org/w/index.php?title=File:Flag_of_South_Korea.svg *License*: unknown *Contributors*: Various

File:Flag of Zimbabwe.svg *Source*: http://en.wikipedia.org/w/index.php?title=File:Flag_of_Zimbabwe.svg *License*: unknown *Contributors*: User:Madden

File:Flag of Mexico.svg *Source*: http://en.wikipedia.org/w/index.php?title=File:Flag_of_Mexico.svg *License*: unknown *Contributors*: User:AlexCovarrubias

File:Flag of Russia.svg *Source*: http://en.wikipedia.org/w/index.php?title=File:Flag_of_Russia.svg *License*: unknown *Contributors*: Anomie

File:Flag of Japan.svg *Source*: http://en.wikipedia.org/w/index.php?title=File:Flag_of_Japan.svg *License*: unknown *Contributors*: Anomie

File:Flag of the People's Republic of China.svg *Source*: http://en.wikipedia.org/w/index.php?title=File:Flag_of_the_People's_Republic_of_China.svg *License*: unknown *Contributors*: User:Denelson83, User:SKopp, User:Shizhao, User:Zscout370

File:Flag of Italy.svg *Source*: http://en.wikipedia.org/w/index.php?title=File:Flag_of_Italy.svg *License*: unknown *Contributors*: Anomie

File:Flag of Estonia.svg *Source*: http://en.wikipedia.org/w/index.php?title=File:Flag_of_Estonia.svg *License*: unknown *Contributors*: User:PeepP, User:SKopp

File:Portland or 1897.png *Source*: http://en.wikipedia.org/w/index.php?title=File:Portland_or_1897.png *License*: unknown *Contributors*: U.S. Geological Survey + EncMstr

File:Portland rose.jpg *Source*: http://en.wikipedia.org/w/index.php?title=File:Portland_rose.jpg *License*: unknown *Contributors*: Stan Shebs, Weezering

File:Portland Max Tunnel.jpg *Source*: http://en.wikipedia.org/w/index.php?title=File:Portland_Max_Tunnel.jpg *License*: unknown *Contributors*: timato

File:pdxspread.jpg *Source*: http://en.wikipedia.org/w/index.php?title=File:Pdxspread.jpg *License*: unknown *Contributors*: User:Infoeco

File:Portland and mt hood.jpg *Source*: http://en.wikipedia.org/w/index.php?title=File:Portland_and_mt_hood.jpg *License*: unknown *Contributors*: User:Jsayre64

File:Mercy corps logo.png *Source*: http://en.wikipedia.org/w/index.php?title=File:Mercy_corps_logo.png *License*: unknown *Contributors*: Gobonobo

File:Mercy Corps headquarters - Portland, Oregon.JPG *Source*: http://en.wikipedia.org/w/index.php?title=File:Mercy_Corps_headquarters_-_Portland,_Oregon.JPG *License*: unknown *Contributors*: User:Aboutmovies

File:PacNW satellite.JPG *Source*: http://en.wikipedia.org/w/index.php?title=File:PacNW_satellite.JPG *License*: unknown *Contributors*: Original uploader was SebastianHelm at en.wikipedia

File:Pacific-Northwest.jpg *Source*: http://en.wikipedia.org/w/index.php?title=File:Pacific-Northwest.jpg *License*: unknown *Contributors*: Badzil, Grendelkhan, JoJan, Kaidor, Mattes, Timeshifter

File:PacNWComparison.PNG *Source*: http://en.wikipedia.org/w/index.php?title=File:PacNWComparison.PNG *License*: unknown *Contributors*: User:SebastianHelm

File:Cascadia megacity map.png *Source*: http://en.wikipedia.org/w/index.php?title=File:Cascadia_megacity_map.png *License*: unknown *Contributors*: Citynoise

File:Qwest seattle sounders pregame.jpg *Source*: http://en.wikipedia.org/w/index.php?title=File:Qwest_seattle_sounders_pregame.jpg *License*: unknown *Contributors*: Luis Antonio Rodríguez Ochoa

Image:Koblenz im Buga-Jahr 2011 - Deutsches Eck 01.jpg *Source*: http://en.wikipedia.org/w/index.php?title=File:Koblenz_im_Buga-Jahr_2011_-_Deutsches_Eck_01.jpg *License*: unknown *Contributors*: Holger Weinandt

Image:Confluence.JPG *Source*: http://en.wikipedia.org/w/index.php?title=File:Confluence.JPG *License*: unknown *Contributors*: Photograph taken by Mark A. Wilson (Department of Geology, The College of Wooster).

Image:IHNCconfluence.png *Source*: http://en.wikipedia.org/w/index.php?title=File:IHNCconfluence.png *License*: unknown *Contributors*: -

Image:Manaus Encontro das aguas 10 2006 103 8x6.jpg *Source*: http://en.wikipedia.org/w/index.php?title=File:Manaus_Encontro_das_aguas_10_2006_103_8x6.jpg *License*: unknown *Contributors*: User:Mariordo

Image: USACE Fremont Bridge Portland.jpg *Source*: http://en.wikipedia.org/w/index.php?title=File:USACE_Fremont_Bridge_Portland.jpg *License*: unknown *Contributors*: Bob Heims, U.S. Army Corps of Engineers

Image: Willametterivermap.jpg *Source*: http://en.wikipedia.org/w/index.php?title=File:Willametterivermap.jpg *License*: unknown *Contributors*: User:Shannon1

File:Speaker Icon.svg *Source*: http://en.wikipedia.org/w/index.php?title=File:Speaker_Icon.svg *License*: unknown *Contributors*: Blast, G.Hagedorn, Mobius, 2 anonymous edits

File:Willamette River.jpg *Source*: http://en.wikipedia.org/w/index.php?title=File:Willamette_River.jpg *License*: unknown *Contributors*: NOAA

File:Erratic rock Oregon.JPG *Source*: http://en.wikipedia.org/w/index.php?title=File:Erratic_rock_Oregon.JPG *License*: unknown *Contributors*: User:Aboutmovies

Image:Willamette River at Molalla SP eb.jpg *Source*: http://en.wikipedia.org/w/index.php?title=File:Willamette_River_at_Molalla_SP_eb.jpg *License*: unknown *Contributors*: Aboutmovies

Image:Oregon City and Willamette Falls, 1867.jpg *Source*: http://en.wikipedia.org/w/index.php?title=File:Oregon_City_and_Willamette_Falls,_1867.jpg *License*: unknown *Contributors*: Howcheng, Jappalang, Mtsmallwood

File:Conveyor belt for Big Pipe project - Portland, Oregon.JPG *Source*: http://en.wikipedia.org/w/index.php?title=File:Conveyor_belt_for_Big_Pipe_project_-_Portland,_Oregon.JPG *License*: unknown *Contributors*: User:Aboutmovies

Image:USACE Detroit Dam Oregon.jpg *Source*: http://en.wikipedia.org/w/index.php?title=File:USACE_Detroit_Dam_Oregon.jpg *License*: unknown *Contributors*: Bob Heims, U.S. Army Corps of Engineers

File:SteelBridgePano1.jpg *Source*: http://en.wikipedia.org/w/index.php?title=File:SteelBridgePano1.jpg *License*: unknown *Contributors*: User:Cacophony

File:Willamette River 1996 flood aerial.jpg *Source*: http://en.wikipedia.org/w/index.php?title=File:Willamette_River_1996_flood_aerial.jpg *License*: unknown *Contributors*: Credit: National Weather Service Portland / Courtesy of U.S. Army Corps of Engineers

File:Willamette River 1996 flooding map.jpg *Source*: http://en.wikipedia.org/w/index.php?title=File:Willamette_River_1996_flooding_map.jpg *License*: unknown *Contributors*: U.S. Army Corps of Engineers

File:FleetWeekRoseFestival.jpg *Source*: http://en.wikipedia.org/w/index.php?title=File:FleetWeekRoseFestival.jpg *License*: unknown *Contributors*: User:Cacophony

File:Osprey mg 9605.jpg *Source*: http://en.wikipedia.org/w/index.php?title=File:Osprey_mg_9605.jpg *License*: unknown *Contributors*: Mike Baird from Morro Bay, USA

File:United States Administrative Divisions Cities.svg *Source*: http://en.wikipedia.org/w/index.php?title=File:United_States_Administrative_Divisions_Cities.svg *License*: unknown *Contributors*: User:Nandhp

File:Manhattan3 amk.jpg *Source*: http://en.wikipedia.org/w/index.php?title=File:Manhattan3_amk.jpg *License*: unknown *Contributors*: user:AngMoKio

File:LA Skyline Mountains2.jpg *Source*: http://en.wikipedia.org/w/index.php?title=File:LA_Skyline_Mountains2.jpg *License*: unknown *Contributors*: User:Nserrano

File:ChicagovanafSearsTower.jpg *Source*: http://en.wikipedia.org/w/index.php?title=File:ChicagovanafSearsTower.jpg *License*: unknown *Contributors*: Original uploader was Voogd075 at nl.wikipedia

File:Panoramic Houston skyline.jpg *Source*: http://en.wikipedia.org/w/index.php?title=File:Panoramic_Houston_skyline.jpg *License*: unknown *Contributors*: User:Hequals2henry

File:Philadelphia skyline from south street bridge.jpg *Source*: http://en.wikipedia.org/w/index.php?title=File:Philadelphia_skyline_from_south_street_bridge.jpg *License*: unknown *Contributors*: Original uploader was Bmoredlj at en.wikipedia

File:Skyline of PHX.jpg *Source*: http://en.wikipedia.org/w/index.php?title=File:Skyline_of_PHX.jpg *License*: unknown *Contributors*: matsubatsu from Tucson, AZ, United States

File:Downtown San Antonio View.JPG *Source*: http://en.wikipedia.org/w/index.php?title=File:Downtown_San_Antonio_View.JPG *License*: unknown *Contributors*: ZeChief

File:San Diego Skyline during winter of Januray 2011.jpg *Source*: http://en.wikipedia.org/w/index.php?title=File:San_Diego_Skyline_during_winter_of_Januray_2011.jpg *License*: unknown *Contributors*: 08OceanBeach SD, FlickreviewR

File:Downtown Dallas from the Trinity River Greenbelt Park, late afternoon, 05-05-11.jpg *Source*: http://en.wikipedia.org/w/index.php?title=File:Downtown_Dallas_from_the_Trinity_River_Greenbelt_Park,_late_afternoon,_05-05-11.jpg *License*: unknown *Contributors*: User:Cordphaeton

File:Downtown San Jose.PNG *Source*: http://en.wikipedia.org/w/index.php?title=File:Downtown_San_Jose.PNG *License*: unknown *Contributors*: I Wiki Edit

File:Jacksonville Skyline Panorama 3.jpg *Source*: http://en.wikipedia.org/w/index.php?title=File:Jacksonville_Skyline_Panorama_3.jpg *License*: unknown *Contributors*: User:digon3

File:Downtown indy from parking garage zoom.JPG *Source*: http://en.wikipedia.org/w/index.php?title=File:Downtown_indy_from_parking_garage_zoom.JPG *License*: unknown *Contributors*: Original uploader was Jasssmit at en.wikipedia

File:FinancialNorth.jpg *Source*: http://en.wikipedia.org/w/index.php?title=File:FinancialNorth.jpg *License*: unknown *Contributors*: Original uploader was Danjahner at en.wikipedia

File:AustinWikiPic2.jpg *Source*: http://en.wikipedia.org/w/index.php?title=File:AustinWikiPic2.jpg *License*: unknown *Contributors*: User:Wikidiculous

File:Columbus-ohio-skyline-panorama.jpg *Source*: http://en.wikipedia.org/w/index.php?title=File:Columbus-ohio-skyline-panorama.jpg *License*: unknown *Contributors*: User:Tysto

File:FortWorthTexasSkylineW.jpg *Source*: http://en.wikipedia.org/w/index.php?title=File:FortWorthTexasSkylineW.jpg *License*: unknown *Contributors*: Original uploader was Gdvanc at en.wikipedia

File:Charlotte Skyline 2011 - Ricky W.jpg *Source*: http://en.wikipedia.org/w/index.php?title=File:Charlotte_Skyline_2011_-_Ricky_W.jpg *License*: unknown *Contributors*: User:Riction

File:DetroitSkyline.jpg *Source*: http://en.wikipedia.org/w/index.php?title=File:DetroitSkyline.jpg *License*: unknown *Contributors*: Shawn Wilson

File:Cityscape of El Paso at Dusk.jpg *Source*: http://en.wikipedia.org/w/index.php?title=File:Cityscape_of_El_Paso_at_Dusk.jpg *License*: unknown *Contributors*: User:Kristiansr

File:Memphis skyline from the air.jpg *Source*: http://en.wikipedia.org/w/index.php?title=File:Memphis_skyline_from_the_air.jpg *License*: unknown *Contributors*: Original uploader was Leonard23 at en.wikipedia

File:Bmore skyline inner harbor.jpg *Source*: http://en.wikipedia.org/w/index.php?title=File:Bmore_skyline_inner_harbor.jpg *License*: unknown *Contributors*: User:Nfutvol

File:DowntownBoston.jpg *Source*: http://en.wikipedia.org/w/index.php?title=File:DowntownBoston.jpg *License*: unknown *Contributors*: LC-de, Lycaon, Mortadelo2005, Urban, Warofdreams

File:Seattle 7.jpg *Source*: http://en.wikipedia.org/w/index.php?title=File:Seattle_7.jpg *License*: unknown *Contributors*: User:Dschwen

File:WashMonument WhiteHouse.jpg *Source*: http://en.wikipedia.org/w/index.php?title=File:WashMonument_WhiteHouse.jpg *License*: unknown *Contributors*: U.S. Air Force Tech. Sgt. Andy Dunaway

File:Nashvilleskyline23.jpg *Source*: http://en.wikipedia.org/w/index.php?title=File:Nashvilleskyline23.jpg *License*: unknown *Contributors*: User:Bsutto04

File:2006-03-26 Denver Skyline I-25 Speer.jpg *Source*: http://en.wikipedia.org/w/index.php?title=File:2006-03-26_Denver_Skyline_I-25_Speer.jpg *License*: unknown *Contributors*: Infrogmation, J 1982, MattWright, Urban, Xnatedawgx

File:Louisville Skyline.jpg *Source*: http://en.wikipedia.org/w/index.php?title=File:Louisville_Skyline.jpg *License*: unknown *Contributors*: Chris Watson

File:Milwaukee skyline.jpg *Source*: http://en.wikipedia.org/w/index.php?title=File:Milwaukee_skyline.jpg *License*: unknown *Contributors*: Original uploader was Miwdke at en.wikipedia

File:Portland_and_Mt_Hood.jpg *Source*: http://en.wikipedia.org/w/index.php?title=File:Portland_and_Mt_Hood.jpg *License*: unknown *Contributors*: User:Amateria1121

File:Downtown Las Vegas from Stratosphere 3.jpg *Source*: http://en.wikipedia.org/w/index.php?title=File:Downtown_Las_Vegas_from_Stratosphere_3.jpg *License*: unknown *Contributors*: Ken Lund from Las Vegas, Nevada, USA

File:Downtown Oklahoma City.jpg *Source*: http://en.wikipedia.org/w/index.php?title=File:Downtown_Oklahoma_City.jpg *License*: unknown *Contributors*: User:Urbanative

File:Alb Skyline.png *Source*: http://en.wikipedia.org/w/index.php?title=File:Alb_Skyline.png *License*: unknown *Contributors*: User:Dschwen, User:JHarrelson

File:CatalinasAndTusconAZ.JPG *Source*: http://en.wikipedia.org/w/index.php?title=File:CatalinasAndTusconAZ.JPG *License*: unknown *Contributors*: User:Howcheng

File:Downtownfresnoskyline.jpg *Source*: http://en.wikipedia.org/w/index.php?title=File:Downtownfresnoskyline.jpg *License*: unknown *Contributors*: 08OceanBeach SD, EugeneZelenko, FieldMarine, Mortadelo2005, Xnatedawgx

File:Sacramento Riverfront.jpg *Source*: http://en.wikipedia.org/w/index.php?title=File:Sacramento_Riverfront.jpg *License*: unknown *Contributors*: Original uploader was DevinCook at en.wikipedia

File:Long Beach, CA at night.jpg *Source*: http://en.wikipedia.org/w/index.php?title=File:Long_Beach,_CA_at_night.jpg *License*: unknown *Contributors*: Original uploader Ed g2s to English Wikipedia

File:Kansas City MO Skyline 14July2008v.jpg *Source*: http://en.wikipedia.org/w/index.php?title=File:Kansas_City_MO_Skyline_14July2008v.jpg *License*: unknown *Contributors*: User:Charvex

File:Downtown Mesa Arizona.jpg *Source*: http://en.wikipedia.org/w/index.php?title=File:Downtown_Mesa_Arizona.jpg *License*: unknown *Contributors*: User:Ixnayonthetimmay

File:VA Beach Town Center.jpg *Source*: http://en.wikipedia.org/w/index.php?title=File:VA_Beach_Town_Center.jpg *License*: unknown *Contributors*: User:Kubigula

File:Atlanta Skyline from Buckhead.jpg *Source*: http://en.wikipedia.org/w/index.php?title=File:Atlanta_Skyline_from_Buckhead.jpg *License*: unknown *Contributors*: Chuck Koehler from Cartersville, GA, USA

File:CC COSPRINGS.jpg *Source*: http://en.wikipedia.org/w/index.php?title=File:CC_COSPRINGS.jpg *License*: unknown *Contributors*: Original uploader was Postoak at en.wikipedia

File:Omaha, Nebraska, USA 2008.jpg *Source*: http://en.wikipedia.org/w/index.php?title=File:Omaha,_Nebraska,_USA_2008.jpg *License*: unknown *Contributors*: Original uploader was JonClee86 at en.wikipedia

File:Downtown-Raleigh-from-Western-Boulevard-Overpass-20081012.jpeg *Source*: http://en.wikipedia.org/w/index.php?title=File:Downtown-Raleigh-from-Western-Boulevard-Overpass-20081012.jpeg *License*: unknown *Contributors*: User:jmturner

File:Downtown Miami skyline May 2011.jpg *Source*: http://en.wikipedia.org/w/index.php?title=File:Downtown_Miami_skyline_May_2011.jpg *License*: unknown *Contributors*: User:Daniel Christensen

File:Cleveland From The Lake.jpg *Source*: http://en.wikipedia.org/w/index.php?title=File:Cleveland_From_The_Lake.jpg *License*: unknown *Contributors*: Original uploader was Aivazovsky at en.wikipedia

File:Tulsa, Oklahoma.jpg *Source*: http://en.wikipedia.org/w/index.php?title=File:Tulsa,_Oklahoma.jpg *License*: unknown *Contributors*: Caleb Long Original uploader was CPacker at en.wikipedia

File:Oaklandatnight02192006.JPG *Source*: http://en.wikipedia.org/w/index.php?title=File:Oaklandatnight02192006.JPG *License*: unknown *Contributors*: User:Miskatonic

File:MinneapolisSkyline.jpg *Source*: http://en.wikipedia.org/w/index.php?title=File:MinneapolisSkyline.jpg *License*: unknown *Contributors*: User:PhotosByJeremiah

File:Wichita-ks.jpg *Source*: http://en.wikipedia.org/w/index.php?title=File:Wichita-ks.jpg *License*: unknown *Contributors*: Original uploader was Nnez4321 at en.wikipedia

File:Cowboysstadium js crop.jpg *Source*: http://en.wikipedia.org/w/index.php?title=File:Cowboysstadium_js_crop.jpg *License*: unknown *Contributors*: User:Beyond My Ken, User:Dopefish

File:Skyline of San Juan, Puerto Rico.jpg *Source*: http://en.wikipedia.org/w/index.php?title=File:Skyline_of_San_Juan,_Puerto_Rico.jpg *License*: unknown *Contributors*: Original uploader was M gonzalez at en.wikipedia

File:Densityhisto.jpg *Source*: http://en.wikipedia.org/w/index.php?title=File:Densityhisto.jpg *License*: unknown *Contributors*: Loodog (talk) Original uploader was Loodog at en.wikipedia

File:United States Administrative Divisions unnumbered.png *Source*: http://en.wikipedia.org/w/index.php?title=File:United_States_Administrative_Divisions_unnumbered.png *License*: unknown *Contributors*: Original uploader was Geuiwogbil at en.wikipedia Later versions were uploaded by Scifiintel at en.wikipedia.

File:Flag of British Columbia.svg *Source*: http://en.wikipedia.org/w/index.php?title=File:Flag_of_British_Columbia.svg *License*: unknown *Contributors*: specification sheets by Christopher Southworth, -xfi-'s source code

File:Coat of Arms of British Columbia.png *Source*: http://en.wikipedia.org/w/index.php?title=File:Coat_of_Arms_of_British_Columbia.png *License*: unknown *Contributors*: unknown. Original uploader was Flrn at en.wikipedia. Later version(s) were uploaded by Noir at en.wikipedia.

File:British Columbia, Canada.svg *Source*: http://en.wikipedia.org/w/index.php?title=File:British_Columbia,_Canada.svg *License*: unknown *Contributors*: User:EOZyo

Image:Speakerlink.svg *Source*: http://en.wikipedia.org/w/index.php?title=File:Speakerlink.svg *License*: unknown *Contributors*: Woodstone. Original uploader was Woodstone at en.wikipedia

File:BC-relief.png *Source*: http://en.wikipedia.org/w/index.php?title=File:BC-relief.png *License*: unknown *Contributors*: Qyd

Image:Strait of Georgia.jpg *Source*: http://en.wikipedia.org/w/index.php?title=File:Strait_of_Georgia.jpg *License*: unknown *Contributors*: Rémih, Sascha Brück, Vmenkov

Image:Mount Robson2.jpg *Source*: http://en.wikipedia.org/w/index.php?title=File:Mount_Robson2.jpg *License*: unknown *Contributors*: Wofratz on German Wikipedia

Image:TakakkawFalls2 edit.jpg *Source*: http://en.wikipedia.org/w/index.php?title=File:TakakkawFalls2_edit.jpg *License*: unknown *Contributors*: User:Fir0002

Image:Cheakamuslake.JPG *Source*: http://en.wikipedia.org/w/index.php?title=File:Cheakamuslake.JPG *License*: unknown *Contributors*: ec

Image:Kwakiutl house pole InvMH975-123-1 .jpg *Source*: http://en.wikipedia.org/w/index.php?title=File:Kwakiutl_house_pole_InvMH975-123-1_.jpg *License*: unknown *Contributors*: User:Jastrow

Image:LastSpike Craigellachie BC Canada.jpg *Source*: http://en.wikipedia.org/w/index.php?title=File:LastSpike_Craigellachie_BC_Canada.jpg *License*: unknown *Contributors*: Ross, Alexander, Best & Co., Winnipeg

Image:Japanese internment camp in British Columbia.jpg *Source*: http://en.wikipedia.org/w/index.php?title=File:Japanese_internment_camp_in_British_Columbia.jpg *License*: unknown *Contributors*: Jack Long

Image:British Columbia Regiment 1940.jpg *Source*: http://en.wikipedia.org/w/index.php?title=File:British_Columbia_Regiment_1940.jpg *License*: unknown *Contributors*: Claude Dettloff, photographer

Image:I 61926.gif *Source*: http://en.wikipedia.org/w/index.php?title=File:I_61926.gif *License*: unknown *Contributors*: Fishhead64

File:Coquihalla Highway.jpg *Source*: http://en.wikipedia.org/w/index.php?title=File:Coquihalla_Highway.jpg *License*: unknown *Contributors*: Colin Keigher at en.wikipedia

Image:Yellowhead-pass sign.jpg *Source*: http://en.wikipedia.org/w/index.php?title=File:Yellowhead-pass_sign.jpg *License*: unknown *Contributors*: Hans-Peter Eckhardt at de.wikipedia

Image:Vancouver ib.jpg *Source*: http://en.wikipedia.org/w/index.php?title=File:Vancouver_ib.jpg *License*: unknown *Contributors*: Thom Quine

File:Vancouver downtown.jpg *Source*: http://en.wikipedia.org/w/index.php?title=File:Vancouver_downtown.jpg *License*: unknown *Contributors*: Bobanny, Oxam Hartog, Themightyquill

Image:AlexFraserBridge.jpg *Source*: http://en.wikipedia.org/w/index.php?title=File:AlexFraserBridge.jpg *License*: unknown *Contributors*: Bobanny

Image:Eastbound over SCB.jpg *Source*: http://en.wikipedia.org/w/index.php?title=File:Eastbound_over_SCB.jpg *License*: unknown *Contributors*: Davepape, Gürbetaler, Iain Bell, Ronaldino, Thryduulf, Voyager, 3 anonymous edits

Image:BC Legislature Buildings and Undersea Gardens.jpg *Source*: http://en.wikipedia.org/w/index.php?title=File:BC_Legislature_Buildings_and_Undersea_Gardens.jpg *License*: unknown *Contributors*: Voyager, 1 anonymous edits

Image:BC-Legislature.jpg *Source*: http://en.wikipedia.org/w/index.php?title=File:BC-Legislature.jpg *License*: unknown *Contributors*: User:Makaristos

Image:Whistler ice sailor.jpg *Source*: http://en.wikipedia.org/w/index.php?title=File:Whistler_ice_sailor.jpg *License*: unknown *Contributors*: Thomas Quine

Image:Cannabis sativa.jpg *Source*: http://en.wikipedia.org/w/index.php?title=File:Cannabis_sativa.jpg *License*: unknown *Contributors*: BLueFiSH.as, D-Kuru, Greatpatton, Hiuppo, Timeshifter, WayneRay, 8 anonymous edits

Image:Portland metro area.png *Source*: http://en.wikipedia.org/w/index.php?title=File:Portland_metro_area.png *License*: unknown *Contributors*: User:Noddy

File:SeattleMontage.png *Source*: http://en.wikipedia.org/w/index.php?title=File:SeattleMontage.png *License*: unknown *Contributors*: Original uploader was Dolphin Jedi at en.wikipedia

File:Flag of Seattle.svg *Source*: http://en.wikipedia.org/w/index.php?title=File:Flag_of_Seattle.svg *License*: unknown *Contributors*: InverseHypercube, Mechamind90, Sfan00 IMG, 1 anonymous edits

File:Seattle seal.png *Source*: http://en.wikipedia.org/w/index.php?title=File:Seattle_seal.png *License*: unknown *Contributors*: Admrboltz, Bobblehead, Cacophony, DangApricot, Gurchzilla, Malo, Neal Finne, Nehrams2020, Shizhao, Shyam, Skier Dude, Stan Shebs, 2 anonymous edits

File:King_County_Washington_Incorporated_and_Unincorporated_areas_Seattle_Highlighted.svg *Source*: http://en.wikipedia.org/w/index.php?title=File:King_County_Washington_Incorporated_and_Unincorporated_areas_Seattle_Highlighted.svg *License*: unknown *Contributors*: Arkyan

File:General history, Alaska Yukon Pacific Exposition, fully illustrated - meet me in Seattle 1909 - Page 70.jpg *Source*: http://en.wikipedia.org/w/index.php?title=File:General_history,_Alaska_Yukon_Pacific_Exposition,_fully_illustrated_-_meet_me_in_Seattle_1909_-_Page_70.jpg *License*: unknown *Contributors*: Peiser / L.B. Franklin, Inc. Scanned by the Seattle Public Library.

File:Seattle - First Ave at Columbia c. 1870.jpg *Source*: http://en.wikipedia.org/w/index.php?title=File:Seattle_-_First_Ave_at_Columbia_c._1870.jpg *License*: unknown *Contributors*: Unknown

File:Seattle - Occidental and Yesler - 1884.jpg *Source*: http://en.wikipedia.org/w/index.php?title=File:Seattle_-_Occidental_and_Yesler_-_1884.jpg *License*: unknown *Contributors*: unattributed

File:Alaska Yukon Pacific Exposition - Rainier Vista.jpg *Source*: http://en.wikipedia.org/w/index.php?title=File:Alaska_Yukon_Pacific_Exposition_-_Rainier_Vista.jpg *License*: unknown *Contributors*: Bjh21, Bobblehead, DrKiernan, Finnrind, Infrogmation, Jmabel, Kevmin, Urban

File:Pioneer square.jpg *Source*: http://en.wikipedia.org/w/index.php?title=File:Pioneer_square.jpg *License*: unknown *Contributors*: Photographer: C.F. Todd ,Sepia: ShadowDragon. Original uploader was ShadowDragon at en.wikipedia

File:Seattle Monorail under construction - 1961.jpg *Source*: http://en.wikipedia.org/w/index.php?title=File:Seattle_Monorail_under_construction_-_1961.jpg *License*: unknown *Contributors*: Seattle City Engineering Department

File:Seattle Ferry.jpg *Source*: http://en.wikipedia.org/w/index.php?title=File:Seattle_Ferry.jpg *License*: unknown *Contributors*: Jamies

File:Space Needle 360 Panorama.jpg *Source*: http://en.wikipedia.org/w/index.php?title=File:Space_Needle_360_Panorama.jpg *License*: unknown *Contributors*: A. Hornung

File:Seattle 07752.JPG *Source*: http://en.wikipedia.org/w/index.php?title=File:Seattle_07752.JPG *License*: unknown *Contributors*: Walter Siegmund

Image:Downtown Seattle 2.JPG *Source*: http://en.wikipedia.org/w/index.php?title=File:Downtown_Seattle_2.JPG *License*: unknown *Contributors*: Foreverprovence, Jmabel, Marku1988

File:Seattle-myst.jpg *Source*: http://en.wikipedia.org/w/index.php?title=File:Seattle-myst.jpg *License*: unknown *Contributors*: User:Cyverius

File:Aerial Downtown Seattle November 2011 - 2.jpg *Source*: http://en.wikipedia.org/w/index.php?title=File:Aerial_Downtown_Seattle_November_2011_-_2.jpg *License*: unknown *Contributors*: User:Jelson25

File:Panorama_Kerry_Park_Seattle.jpg *Source*: http://en.wikipedia.org/w/index.php?title=File:Panorama_Kerry_Park_Seattle.jpg *License*: unknown *Contributors*: Westonmr

File:Seattle 3.jpg *Source*: http://en.wikipedia.org/w/index.php?title=File:Seattle_3.jpg *License*: unknown *Contributors*: User:Dschwen

File:SCL.jpg *Source*: http://en.wikipedia.org/w/index.php?title=File:SCL.jpg *License*: unknown *Contributors*: DVD R W

File:Pike Place Market Seattle.jpg *Source*: http://en.wikipedia.org/w/index.php?title=File:Pike_Place_Market_Seattle.jpg *License*: unknown *Contributors*: User:Dschwen

File:Seattle - The Moore Theater entrance 01.jpg *Source*: http://en.wikipedia.org/w/index.php?title=File:Seattle_-_The_Moore_Theater_entrance_01.jpg *License*: unknown *Contributors*: User:Jmabel

File:Seattle Cruise Ship.jpg *Source*: http://en.wikipedia.org/w/index.php?title=File:Seattle_Cruise_Ship.jpg *License*: unknown *Contributors*: mahalie from seattle, USA

File:Seattle Sheraton from Two Union Square.jpg *Source*: http://en.wikipedia.org/w/index.php?title=File:Seattle_Sheraton_from_Two_Union_Square.jpg *License*: unknown *Contributors*: User:Cumulus Clouds

File:Lake Union Plane.jpg *Source*: http://en.wikipedia.org/w/index.php?title=File:Lake_Union_Plane.jpg *License*: unknown *Contributors*: User:Bubsty

File:SafecoFld.JPG *Source*: http://en.wikipedia.org/w/index.php?title=File:SafecoFld.JPG *License*: unknown *Contributors*: Galaksiafervojo, KAMiKAZOW, 2 anonymous edits

File:Qwest Field Nighttime.jpg *Source*: http://en.wikipedia.org/w/index.php?title=File:Qwest_Field_Nighttime.jpg *License*: unknown *Contributors*: Original uploader was Rattlhed at en.wikipedia

File:Greenlaketrail.jpg *Source*: http://en.wikipedia.org/w/index.php?title=File:Greenlaketrail.jpg *License*: unknown *Contributors*: Original uploader was Popcultureguy at en.wikipedia

File:Wamuseattle.jpg *Source*: http://en.wikipedia.org/w/index.php?title=File:Wamuseattle.jpg *License*: unknown *Contributors*: Jmabel, Superm401, Venske, 3 anonymous edits

File:Seattle City Hall 01.jpg *Source*: http://en.wikipedia.org/w/index.php?title=File:Seattle_City_Hall_01.jpg *License*: unknown *Contributors*: User:Jmabel

File:University of Washington Quad, Spring 2007.jpg *Source*: http://en.wikipedia.org/w/index.php?title=File:University_of_Washington_Quad,_Spring_2007.jpg *License*: unknown *Contributors*: User:Punctured Bicycle

File:SeattleI5Skyline.jpg *Source*: http://en.wikipedia.org/w/index.php?title=File:SeattleI5Skyline.jpg *License*: unknown *Contributors*: User:Cacophony

File:Link Tunell Sound transit.jpg *Source*: http://en.wikipedia.org/w/index.php?title=File:Link_Tunell_Sound_transit.jpg *License*: unknown *Contributors*: User:Fanboy

File:New Seattle Steam plant Western Avenue downtown Seattle Washington.JPG *Source*: http://en.wikipedia.org/w/index.php?title=File:New_Seattle_Steam_plant_Western_Avenue_downtown_Seattle_Washington.JPG *License*: unknown *Contributors*: User:MarmadukePercy

File:Flag of Norway.svg *Source*: http://en.wikipedia.org/w/index.php?title=File:Flag_of_Norway.svg *License*: unknown *Contributors*: User:Dbenbenn

File:Flag of the Philippines.svg *Source*: http://en.wikipedia.org/w/index.php?title=File:Flag_of_the_Philippines.svg *License*: unknown *Contributors*: Aira Cutamora

File:Flag of New Zealand.svg *Source*: http://en.wikipedia.org/w/index.php?title=File:Flag_of_New_Zealand.svg *License*: unknown *Contributors*: Adabow, Adambro, Arria Belli, Avenue, Bawolff, Bjankuloski06en, ButterStick, Denelson83, Donk, Duduziq, EugeneZelenko, Fred J, Fry1989, Hugh Jass, Ibagli, Jusjih, Klemen Kocjancic, Mamndassan, Mattes, Nightstallion, O, Peeperman, Poromiami, Reisio, Rfc1394, Shizhao, Tabasco, Transparent Blue, Väsk, Xufanc, Zscout370, 35 anonymous edits

File:Flag of Ireland.svg *Source*: http://en.wikipedia.org/w/index.php?title=File:Flag_of_Ireland.svg *License*: unknown *Contributors*: User:SKopp

File:Flag of Poland.svg *Source*: http://en.wikipedia.org/w/index.php?title=File:Flag_of_Poland.svg *License*: unknown *Contributors*: Anomie, Mifter

File:Flag of Vietnam.svg *Source*: http://en.wikipedia.org/w/index.php?title=File:Flag_of_Vietnam.svg *License*: unknown *Contributors*: user:Lưu Ly

File:Flag of Cameroon.svg *Source*: http://en.wikipedia.org/w/index.php?title=File:Flag_of_Cameroon.svg *License*: unknown *Contributors*: (of code)

File:Flag of Kenya.svg *Source*: http://en.wikipedia.org/w/index.php?title=File:Flag_of_Kenya.svg *License*: unknown *Contributors*: User:Pumbaa80

File:Flag of France.svg *Source*: http://en.wikipedia.org/w/index.php?title=File:Flag_of_France.svg *License*: unknown *Contributors*: Anomie

File:Flag of Hungary.svg *Source*: http://en.wikipedia.org/w/index.php?title=File:Flag_of_Hungary.svg *License*: unknown *Contributors*: User:SKopp

File:Flag of Iceland.svg *Source*: http://en.wikipedia.org/w/index.php?title=File:Flag_of_Iceland.svg *License*: unknown *Contributors*: User:Zscout370, User:Ævar Arnfjörð Bjarmason

File:Flag of Cambodia.svg *Source*: http://en.wikipedia.org/w/index.php?title=File:Flag_of_Cambodia.svg *License*: unknown *Contributors*: User:Nightstallion

File:Flag of Indonesia.svg *Source*: http://en.wikipedia.org/w/index.php?title=File:Flag_of_Indonesia.svg *License*: unknown *Contributors*: User:Gabbe, User:SKopp

File:Flag of Uzbekistan.svg *Source*: http://en.wikipedia.org/w/index.php?title=File:Flag_of_Uzbekistan.svg *License*: unknown *Contributors*: User:Zscout370

File:Cedar Mill Community Library - Oregon.JPG *Source*: http://en.wikipedia.org/w/index.php?title=File:Cedar_Mill_Community_Library_-_Oregon.JPG *License*: unknown *Contributors*: User:Aboutmovies

File:Washington_County_Oregon_Incorporated_and_Unincorporated_areas_Cedar_Mill_Highlighted.svg *Source*: http://en.wikipedia.org/w/index.php?title=File:Washington_County_Oregon_Incorporated_and_Unincorporated_areas_Cedar_Mill_Highlighted.svg *License*: unknown *Contributors*: Arkyan

File:Map of Oregon highlighting Washington County.svg *Source*: http://en.wikipedia.org/w/index.php?title=File:Map_of_Oregon_highlighting_Washington_County.svg *License*: unknown *Contributors*: User:Dbenbenn

File:Map of USA OR.svg *Source*: http://en.wikipedia.org/w/index.php?title=File:Map_of_USA_OR.svg *License*: unknown *Contributors*: User:Huebi

File:Washington County Courthouse front - Hillsboro, Oregon.JPG *Source*: http://en.wikipedia.org/w/index.php?title=File:Washington_County_Courthouse_front_-_Hillsboro,_Oregon.JPG *License*: unknown *Contributors*: Original uploader was Aboutmovies at en.wikipedia

File:Washington County, Oregon map.gif *Source*: http://en.wikipedia.org/w/index.php?title=File:Washington_County,_Oregon_map.gif *License*: unknown *Contributors*: U.S. Federal government

File:I-5.svg *Source*: http://en.wikipedia.org/w/index.php?title=File:I-5.svg *License*: unknown *Contributors*: Augiasstallputzer, Juliancolton, Kazuya35, Ltljltlj, Rocket000, SPUI, 1 anonymous edits

File:I-205.svg *Source*: http://en.wikipedia.org/w/index.php?title=File:I-205.svg *License*: unknown *Contributors*: Ltljltlj, Raquel350, SPUI

File:US 26.svg *Source*: http://en.wikipedia.org/w/index.php?title=File:US_26.svg *License*: unknown *Contributors*: Bidgee, SPUI, 3 anonymous edits

File:OR 6.svg *Source*: http://en.wikipedia.org/w/index.php?title=File:OR_6.svg *License*: unknown *Contributors*: User:Northenglish

File:OR 8.svg *Source*: http://en.wikipedia.org/w/index.php?title=File:OR_8.svg *License*: unknown *Contributors*: User:Northenglish

File:OR 10.svg *Source*: http://en.wikipedia.org/w/index.php?title=File:OR_10.svg *License*: unknown *Contributors*: User:Northenglish

File:OR 47.svg *Source*: http://en.wikipedia.org/w/index.php?title=File:OR_47.svg *License*: unknown *Contributors*: User:Northenglish

File:OR 99E.svg *Source*: http://en.wikipedia.org/w/index.php?title=File:OR_99E.svg *License*: unknown *Contributors*: User:Northenglish

File:OR 210.svg *Source*: http://en.wikipedia.org/w/index.php?title=File:OR_210.svg *License*: unknown *Contributors*: User:Northenglish

File:OR 217.svg *Source*: http://en.wikipedia.org/w/index.php?title=File:OR_217.svg *License*: unknown *Contributors*: User:Northenglish

File:OR 219.svg *Source*: http://en.wikipedia.org/w/index.php?title=File:OR_219.svg *License*: unknown *Contributors*: User:Northenglish

File:Oregon Population Growth by County.png *Source*: http://en.wikipedia.org/w/index.php?title=File:Oregon_Population_Growth_by_County.png *License*: unknown *Contributors*: User:Northwest-historian

File:Washington County Jail - Hillsboro, Oregon.JPG *Source*: http://en.wikipedia.org/w/index.php?title=File:Washington_County_Jail_-_Hillsboro,_Oregon.JPG *License*: unknown *Contributors*: Original uploader was Aboutmovies at en.wikipedia

File:Downtown Beaverton Oregon.jpg *Source*: http://en.wikipedia.org/w/index.php?title=File:Downtown_Beaverton_Oregon.jpg *License*: unknown *Contributors*: Ajbenj at en.wikipedia

File:Forest Grove Oregon Pacific Avenue 1920.jpg *Source*: http://en.wikipedia.org/w/index.php?title=File:Forest_Grove_Oregon_Pacific_Avenue_1920.jpg *License*: unknown *Contributors*: Aboutmovies

File:Lakewood Bay Oswego Lake.jpg *Source*: http://en.wikipedia.org/w/index.php?title=File:Lakewood_Bay_Oswego_Lake.jpg *License*: unknown *Contributors*: User:Esprqii

File:LO flag.jpg *Source*: http://en.wikipedia.org/w/index.php?title=File:LO_flag.jpg *License*: unknown *Contributors*: Ulmanor

File:Seal of lake oswego Oregon.jpg *Source*: http://en.wikipedia.org/w/index.php?title=File:Seal_of_lake_oswego_Oregon.jpg *License*: unknown *Contributors*: Ajbenj

File:Clackamas_County_Oregon_Incorporated_and_Unincorporated_areas_Lake_Oswego_Highlighted.svg *Source*: http://en.wikipedia.org/w/index.php?title=File:Clackamas_County_Oregon_Incorporated_and_Unincorporated_areas_Lake_Oswego_Highlighted.svg *License*: unknown *Contributors*: Arkyan

File:Loudspeaker.svg *Source*: http://en.wikipedia.org/w/index.php?title=File:Loudspeaker.svg *License*: unknown *Contributors*: Bayo, Gmaxwell, Husky, Iamunknown, Mirithing, Myself488, Nethac DIU, Omegatron, Rocket000, The Evil IP address, Wouterhagens, 18 anonymous edits

File:Oregon Iron Company Furnace restored.JPG *Source*: http://en.wikipedia.org/w/index.php?title=File:Oregon_Iron_Company_Furnace_restored.JPG *License*: unknown *Contributors*: User:Esprqii

File:Lake Oswego Railroad Bridge.jpg *Source*: http://en.wikipedia.org/w/index.php?title=File:Lake_Oswego_Railroad_Bridge.jpg *License*: unknown *Contributors*: User:Esprqii

File:Lake Oswego High School.jpg *Source*: http://en.wikipedia.org/w/index.php?title=File:Lake_Oswego_High_School.jpg *License*: unknown *Contributors*: Original uploader was Mfrk at en.wikipedia

File:Lakeridge High School.jpg *Source*: http://en.wikipedia.org/w/index.php?title=File:Lakeridge_High_School.jpg *License*: unknown *Contributors*: Original uploader was Bentoman at en.wikipedia

MIX
Papier aus verantwortungsvollen Quellen
Paper from responsible sources
FSC® C105338

Printed by Books on Demand GmbH, Norderstedt / Germany